iOS Development  sh®

Your visual blueprint for developing Apple® apps

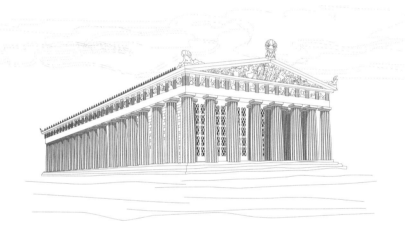

by Julian Dolce

Wiley Publishing, Inc.

iOS Development with Flash®: Your visual blueprint™ for developing Apple® apps

Published by
Wiley Publishing, Inc.
10475 Crosspoint Boulevard
Indianapolis, IN 46256
www.wiley.com

Published simultaneously in Canada

Copyright © 2011 by Wiley Publishing, Inc., Indianapolis, Indiana

No part of this publication may be reproduced, stored in a retrieval system or transmitted in any form or by any means, electronic, mechanical, photocopying, recording, scanning or otherwise, except as permitted under Sections 107 or 108 of the 1976 United States Copyright Act, without either the prior written permission of the Publisher, or authorization through payment of the appropriate per-copy fee to the Copyright Clearance Center, 222 Rosewood Drive, Danvers, MA 01923, (978) 750-8400, fax (978) 646-8600. Requests to the Publisher for permission should be addressed to the Permissions Department, John Wiley & Sons, Inc., 111 River Street, Hoboken, NJ 07030, 201-748-6011, fax 201-748-6008, or online at http://www.wiley.com/go/permissions.

Library of Congress Control Number: 2010939961

ISBN: 978-0-470-62204-9

Manufactured in the United States of America

10 9 8 7 6 5 4 3 2 1

Trademark Acknowledgments

Wiley, the Wiley Publishing logo, Visual, the Visual logo, Visual Blueprint, Read Less - Learn More, and related trade dress are trademarks or registered trademarks of John Wiley & Sons, Inc. and/or its affiliates. IOS is a trademark or registered trademark of Cisco in the U.S. and other countries, and iOS is used under license by Apple. Apple is a registered trademark of Apple Inc. Adobe and Flash are either registered trademarks or trademarks of Adobe Systems Incorporated in the United States and/or other countries. All other trademarks are the property of their respective owners. Wiley Publishing, Inc. is not associated with any product or vendor mentioned in this book.

LIMIT OF LIABILITY/DISCLAIMER OF WARRANTY: THE PUBLISHER AND THE AUTHOR MAKE NO REPRESENTATIONS OR WARRANTIES WITH RESPECT TO THE ACCURACY OR COMPLETENESS OF THE CONTENTS OF THIS WORK AND SPECIFICALLY DISCLAIM ALL WARRANTIES, INCLUDING WITHOUT LIMITATION WARRANTIES OF FITNESS FOR A PARTICULAR PURPOSE. NO WARRANTY MAY BE CREATED OR EXTENDED BY SALES OR PROMOTIONAL MATERIALS. THE ADVICE AND STRATEGIES CONTAINED HEREIN MAY NOT BE SUITABLE FOR EVERY SITUATION. THIS WORK IS SOLD WITH THE UNDERSTANDING THAT THE PUBLISHER IS NOT ENGAGED IN RENDERING LEGAL, ACCOUNTING, OR OTHER PROFESSIONAL SERVICES. IF PROFESSIONAL ASSISTANCE IS REQUIRED, THE SERVICES OF A COMPETENT PROFESSIONAL PERSON SHOULD BE SOUGHT. NEITHER THE PUBLISHER NOR THE AUTHOR SHALL BE LIABLE FOR DAMAGES ARISING HEREFROM. THE FACT THAT AN ORGANIZATION OR WEBSITE IS REFERRED TO IN THIS WORK AS A CITATION AND/OR A POTENTIAL SOURCE OF FURTHER INFORMATION DOES NOT MEAN THAT THE AUTHOR OR THE PUBLISHER ENDORSES THE INFORMATION THE ORGANIZATION OR WEBSITE MAY PROVIDE OR RECOMMENDATIONS IT MAY MAKE. FURTHER, READERS SHOULD BE AWARE THAT INTERNET WEBSITES LISTED IN THIS WORK MAY HAVE CHANGED OR DISAPPEARED BETWEEN WHEN THIS WORK WAS WRITTEN AND WHEN IT IS READ.

FOR PURPOSES OF ILLUSTRATING THE CONCEPTS AND TECHNIQUES DESCRIBED IN THIS BOOK, THE AUTHOR HAS CREATED VARIOUS NAMES, COMPANY NAMES, MAILING, E-MAIL AND INTERNET ADDRESSES, PHONE AND FAX NUMBERS AND SIMILAR INFORMATION, ALL OF WHICH ARE FICTITIOUS. ANY RESEMBLANCE OF THESE FICTITIOUS NAMES, ADDRESSES, PHONE AND FAX NUMBERS AND SIMILAR INFORMATION TO ANY ACTUAL PERSON, COMPANY AND/OR ORGANIZATION IS UNINTENTIONAL AND PURELY COINCIDENTAL.

Contact Us

For general information on our other products and services, please contact our Customer Care Department within the U.S. at 877-762-2974, outside the U.S. at 317-572-3993 or fax 317-572-4002.

For technical support, please visit www.wiley.com/techsupport.

The Parthenon

When work began on the Parthenon in 447 B.C., the Athenian Empire was at the height of its power. Created by architects Iktinos and Kallikrates, the Parthenon served the primary function of sheltering a massive gold and ivory statue of the goddess Athena, created by Pheidias. The stylistic conventions of this magnificent temple have become the paradigm of Classical architecture — a style that has influenced architecture for centuries. Among its subtle and unique architectural features is the fact that the Parthenon is built without a single absolutely straight line.

Discover more about the Parthenon and other ancient Greek temples in *Frommer's Greece* (ISBN 978-0-470-52663-7), available wherever books are sold or at www.Frommers.com.

WILEY

Sales

Contact Wiley
at (877) 762-2974
or (317) 572-4002.

Credits

Acquisitions Editor
Aaron Black

Project Editor
Dana Rhodes Lesh

Technical Editor
Paul Geyer

Copy Editor
Dana Rhodes Lesh

Editorial Director
Robyn Siesky

Editorial Manager
Rosemarie Graham

Business Manager
Amy Knies

Senior Marketing Manager
Sandy Smith

Vice President and Executive Group Publisher
Richard Swadley

Vice President and Executive Publisher
Barry Pruett

Project Coordinator
Katie Crocker

Graphics and Production Specialists
Carrie Cesavice
Joyce Haughey
Andrea Hornberger
Jennifer Mayberry

Quality Control Technician
Laura Albert

Proofreading and Indexing
Jacqui Brownstein
Potomac Indexing, LLC

Media Development Project Manager
Laura Moss

Media Development Assistant Project Manager
Jenny Swisher

Media Development Associate Producers
Josh Frank
Marilyn Hummel
Doug Kuhn
Shawn Patrick

Screen Artists
Ana Carillo
Cheryl Grubbs
Jill A. Proll
Ronald Terry

Cover Art Illustrator
Cheryl Grubbs

About the Author

Julian Dolce is a senior Flash developer at QNX Software Systems, specializing in mobile AIR applications. Julian has spoken at numerous conferences around the world where he has taught workshops on moving from Flash development to iPhone development and on AIR for Android development. He also maintains a personal development blog, www.deleteaso.com, in which he writes about his life as a Flash developer.

Author's Acknowledgments

For Steve Jobs and Willy.

How to Use This Book

Who This Book Is For
This book is for intermediate-to-advanced Flash developers who want to use their knowledge of Flash and ActionScript to develop iPhone applications.

The Conventions in This Book

❶ Steps
This book uses a step-by-step format to guide you easily through each task. Numbered steps are actions you must do; bulleted steps clarify a point, step, or optional feature; and indented steps give you the result.

❷ Notes
Notes give additional information — special conditions that may occur during an operation, a situation that you want to avoid, or a cross-reference to a related area of the book.

❸ Extra or Apply It
An Extra section provides additional information about the preceding task — insider information and tips for ease and efficiency. An Apply It section takes the code from the preceding task one step further and allows you to take full advantage of it.

❹ Bold
Bold type shows text or numbers you must type.

❺ Italics
Italic type introduces and defines a new term.

❻ Courier Font
`Courier font` indicates the use of scripting language code such as statements, operators, or functions, and code such as objects, methods, or properties.

Web Site
You can find the code samples throughout the book on the Wiley Web page for the book, www.wiley.com/go/iosappsvisualblueprint, on the Downloads tab.

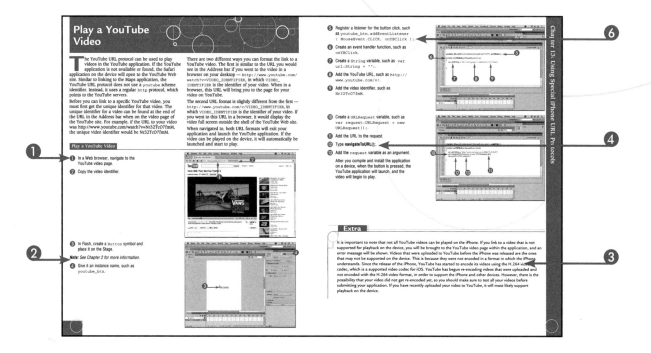

TABLE OF CONTENTS

1 GETTING STARTED WITH IPHONE DEVELOPMENT 2

Introducing the iPhone and iPod touch 2
Introducing the Development Tools 4
Introducing the Available APIs 6
Check What APIs Are Not Available 8
Become an iPhone Developer 10
Generate a Certificate Signing Request on Mac OS X 12
Generate a Certificate Signing Request in Windows 14
Upload Your Certificate Signing Request 16
Create a P12 Certificate on Mac OS X 18
Create a P12 Certificate in Windows 20
Add Devices to Your Account 22
Create App IDs 24
Create a Provisioning Profile 26
Add Provisioning Files to Your Device 28

2 GETTING STARTED WITH FLASH CS5 30

Using the Actions Panel 30
Create a Skeleton Custom Class 32
Set the Source Path 34
Create MovieClips 36
Create Buttons 38
Edit Properties in Flash 40
Add Objects to the Stage with Code 42
Remove Objects from the Stage with Code 44
Work with Events 46
Using the Drawing API 48
Using Flash CS5 Help 50

3 DEVELOPING YOUR FIRST APPLICATION 52

Create a New Project 52
Configure Publish Settings 54
Set Your Application Output 56
Add Your iPhone Certificate 58
Add Your Provisioning File 60
Compile from Flash Professional CS5 62
Compile from the Command Line 64

Install Your Application with iTunes..66
Install Your Application with the iPhone Configuration Utility............................68
Install Your Application with Xcode..70
Update Your Version Number 72

4 DESIGNING YOUR APPLICATION............... 74

Explore Apple's Human Interface Guidelines 74
Understanding Screen Resolutions..76
Create Full-Screen Applications..78
Understanding Screen Orientation 80
Create Usable Hit States..82
Understanding Layout 84
Change the Status Bar Style..86

5 HANDLING INTERACTION 88

Create Button States..88
Respond to Touch Events 90
Track Multiple Touches..92
Respond to Zoom Events 94
Respond to Rotate Events..96
Respond to Pan Events 98
Respond to Swipe Events...100
Listen for Accelerometer Events..102
Determine If the Accelerometer Is Available 104
Determine Device Orientation 106
Detect Which Way Is Up 108
Filter Accelerometer Data 110

6 WORKING WITH IMAGES 112

Prepare Your Images..112
Import Images..114
Display Images 116
Bundle Images with Your Application..118
Load Images at Runtime 120
Create Images Dynamically 122
Save Images to the Photo Library 124
Load Images from the Photo Library...126
Using iOS Default Images 128

TABLE OF CONTENTS

7 WORKING WITH SOUND 130

Import Audio into Your Project 130
Choose an Audio Codec 132
Bundle Sounds with Your Application 134
Load Sounds at Runtime..136
Play Sounds...138
Stop Sounds...140
Set the Volume of a Sound 142
Visualize the Sound Spectrum 144

8 WORKING WITH VIDEO 146

Explore Available Video Formats and Encode a Video File 146
Convert Videos...148
Embed a Video...150
Bundle a Video with Your Application 152
Load a Video..154
Buffer a Video...156
Control a Video 158
Set the Volume of a Video..160

9 WORKING WITH TEXT 162

Determine Available Fonts on Your Device 162
Embed Fonts in Your Application 164
Create an Input TextField...166
Create a Password TextField 168
Using TLF TextFields 170
Create a Scrollable TextField 172

10 SAVING STATE. 174

Create a Local SharedObject 174
Write to a SharedObject 176
Load Data from a SharedObject 178
Connect to a SQLite Database 180
Create a SQLite Table..182
Insert Data into a SQLite Table 184
Select Data from a SQLite Table 186
Update Data in a SQLite Table 188

Delete Data from a SQLite Table 190
Handle Application Exits 192
Save Application States 194

11 WORKING WITH FILES 196

Reference Files and Directories 196
Write Files 198
Read Files 200
Update Files 202
Append Files 204
Handle Files Synchronously 206
Copy Files 208

12 USING THE LOCATION, CONTACTS, AND WIFI FEATURES 210

Retrieve Your Current Location 210
Map Your Location 212
Determine Your Speed 216
Retrieve a List of Contacts 218
Retrieve a Contact's Details 220
Retrieve Phone Number Favorites 222
Check for an Internet Connection 224
Check for a Persistent WiFi Connection 226
Set the System Idle Mode 228

13 USING SPECIAL IPHONE URL PROTOCOLS 230

Make Phone Calls 230
Open the Mail Application 232
Open the Maps Application 234
Open the Messages Application 236
Play a YouTube Video 238
Open the iTunes Store 240

14 INTEGRATING WITH THIRD-PARTY SERVICES 242

Submit Updates to Twitter 242
Display Ads with Smaato 244

TABLE OF CONTENTS

Track with Google Analytics 246
Display Ads with AdMob 248

15 OPTIMIZING PERFORMANCE............... 250

Optimize Your Display List 250
Manage Mouse Events 252
Understanding cacheAsBitmap 254
Understanding cacheAsBitmapMatrix 256
Determine the Device OS 258

16 CREATING APPLICATION SETTINGS 260

Create a Settings Bundle 260
Add a Settings Group 262
Add a Text Setting 264
Add a Multiple Value Settings Field 266
Add a Toggle Switch Field 268
Add a Slider Settings Field 270
Add a Title Settings Field 272
Add the Settings Bundle to Your Application 274
Read the Settings 276
Add an Icon to Your Settings 278
Read Your Device's Global Settings 280

17 DEBUGGING YOUR APPLICATION 282

Show Your Trace Statements 282
Create Breakpoints 284
Using the Flash CS5 Debugger 286
Understanding the Debug Console 288
Understanding the Variables Panel 290
Get Crash Reports 292
Using Instruments 294
Using the ObjectAlloc Instrument 296
Using the Core Animation Instrument 298
Using the OpenGL ES Instrument 300
Using the Activity Monitor Instrument 302
Using the Hardware Acceleration Profiler 304

18 DEPLOYING YOUR APPLICATION 306

Create an Application Icon 306
Remove the Glare from Your Application Icon 308
Create a Default Splash Screen...310
Create App Store Graphics 312
Create a Distribution Certificate...314
Create an Ad Hoc Provisioning File..316
Publish for Ad Hoc Distribution..318
Create an App Store Provisioning File..320
Publish Your Application for App Store Distribution.......................322
Submit Your Application to the App Store 324
Getting Your App Approved 328
Track Your Application Sales 330

INDEX. 332

Introducing the iPhone and iPod touch

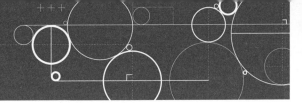

It is an exciting time to be a Flash developer. Adobe has taken big steps in making the Flash Platform available on as many devices as possible. The Open Screen Project is an Adobe-led initiative whose goal is to bring rich Internet experiences seamlessly across as many devices as possible. Flash Player 10.1 will be available for multiple mobile platforms, such as Google Android, RIM's BlackBerry, Palm Pre, and Nokia, as well as numerous other devices such as TVs, set top boxes, tablets, and netbooks. Adobe is working with these and over 50 other partners to optimize Flash Player 10.1 in order to work better with the different devices.

However, one manufacturer that is not part of this initiative is Apple. To allow Flash developers to create iPhone applications, Adobe has created a cross-compiler for the iPhone called the *iPhone Packager*. The iPhone Packager, which comes with Flash CS5, takes your .swf file and converts it to a native iPhone application.

The workflow is similar to what you are familiar with when building your Flash applications for the Web. You write ActionScript 3 code and compile it to an .swf file. An .swf file contains an abc block, which stands for *ActionScript byte code.* This is what all of your ActionScript 3 code gets compiled to when publishing your file. The iPhone Packager goes through your .swf file, finds all of the abc blocks, strips them out the file, and converts them into native ARM assembly code using the LLVM (low level virtual machine) compiler infrastructure. Your application is then signed using the Apple signing process. After it is signed, it is ready to be deployed on a device running iOS (iPhone OS).

There are three devices that run iOS — the iPhone, the iPod touch, and the iPad. Each device has different features and different hardware. It is important to understand the difference between each device. It is a great idea to have as many of the different versions of the devices on hand so that you can test your application on each of them.

iPhone and iPhone 3G

The hardware for the first generation iPhone and the iPhone 3G are very similar. They both have a 620MHz underclocked to 412MHz Samsung 32-bit RISC ARM processor, PowerVR MBX Lite 3D GPU, 128MB DRAM, WiFi, Bluetooth 2.0, and a 2.0 megapixel camera. The biggest difference between the two is their design. The first generation came with an aluminum back, which was later switched to a glossy plastic back for the 3G version. Also, the iPhone 3G has assisted GPS and can communicate over the 3G cell network.

iPhone 3GS

The third generation of the iPhone is known as the 3GS. Released a year after the iPhone 3G, the 3GS saw many improvements and is the fastest of the three generations. With an 833MHz underlocked to 600MHz processor, PowerVR SGX GPU, and 256MB DRAM, you will notice a significant difference in performance compared to the earlier version. Applications will be more responsive, and animations will be a lot smoother. Along with the hardware upgrade, the 3GS also introduced voice control, a digital compass, Nike+, and a 3.0 megapixel camera, which also shoots video in VGA at 30 frames per second.

iPod

The first and second generation iPod touches have very similar hardware to the iPhone and iPhone 3G. The second generation processor ran at 532MHz, up from the earlier 412MHz. Theses iPod touches were made in 8GB, 16GB, and 32GB models, all of which have been discontinued except for the second generation 8GB. In late 2009, Apple released new 32GB and 64GB versions with improved hardware. These versions put the processor, graphics, and memory on par with that of the iPhone 3GS. They also came with earphones with a remote and microphone integrated in. This enables you to interact with the device through the microphone, just like you can with any of the iPhone versions.

iPad

The iPad is Apple's fastest mobile device to date and is almost twice as fast as the latest iPhone and iPod models. With the new 1GHz Apple A4 processor and 512MB DRAM, you can be assured that your applications will feel faster than ever. The biggest change is the 1024 x 768 resolution screen. This is quite a big difference from the 320 x 480 resolution that the iPhone and iPod touch have. Apps created for the iPod and iPhone will still work on the iPad; however, to really take advantage of the platform, you will want to carefully plan your applications to support both.

Test for Multiple Devices

As you can see, there are many differences between all the different models, and the iPad is a game changer. If you want to get a few different devices for testing, you can probably get away with two different devices. If you already have an iPhone, try and get your hands on the iPod with the opposite specs. For example, if you have an iPhone 3GS, pick up one of the second generation 8GB iPod touches before they are discontinued. There are lots of tips and things to think about in this book when developing your applications to support multiple platforms. It is a good exercise to try and take all of these into consideration early on in development. Some things to ask yourself are, "What does my application look like on multiple resolutions?," and "How does a user interact with my application on a nontouch-enabled device?." Even if you ever only plan to support one device today, allowing for multiple platforms in the future will prove to be worthwhile.

Introducing the Development Tools

If you have developed Flash applications before, you will already be familiar with some of the tools that will be explored throughout this book. However, there are a number of new applications that are specific to iPhone development that you may not be so familiar with. Some of these are completely optional when developing iPhone applications, and some are only available on Mac OS X.

Flash CS5 Professional

Flash CS5 is the main integrated development environment (IDE) for developing Flash applications for the Web, desktop, and Flash Lite–enabled mobile devices. In this 11th version of Flash, Flash CS5 introduces us to the ability to publish Flash applications to native iPhone applications. This is a pretty big accomplishment as it brings iPhone development to not only the many Flash developers in the world, but also the Windows operating system. Flash will be the primary application used throughout the book as you explore creating iPhone applications.

Flash Builder

Flash Builder, formerly know as *Flex Builder*, is an Eclipse-based IDE for creating Flex and AS3 projects. Flash Builder is Adobe's main ActionScript coding application. Flash CS5 does have the ability to write separate ActionScript code and classes; however, Flash Builder provides a much more feature-rich development environment. With the newest version of Flash Builder, Adobe has also integrated better workflow between it and Flash CS5. You are now able to publish .fla files directly from Flash Builder without having to switch between applications. One of the benefits of Flash Builder is that it is built on top of Eclipse, a popular open source IDE. This enables you to take advantage of the many plug-ins built for Eclipse, which provide additional functionality that you do not get in Flash CS5. There are many plug-ins for managing source control, build integration, and support for other programming languages.

Xcode

Xcode is part of Apple's developer tools, which can be downloaded with the iPhone SDK from the iPhone Developer Program Portal, http://developer.apple.com/iphone/. Xcode is a Mac OS X–only full-featured development environment for creating applications with the iPhone SDK. It is the primary development tool if you are building Cocoa-based applications. Xcode's Organizer window has some great features to help manage data on your device. You can install and uninstall applications and provisioning files, as well retrieve application data and crash reports from your device. This book will also explore creating application settings bundles with Xcode, which allow you to create application settings views that will appear in the Settings application on your device.

Instruments

Instruments, also part of Apple's developer tools, is an OS X–only application that allows you to profile many different parts of your application. It enables you to collect a wide variety of data at the same time, allowing you to easily compare the data and spot any abnormalities. Instruments uses different instruments to collect data. There is an instrument to profile almost every part of your application. You can profile file access, memory usage, object allocation, and OpenGL. Instruments can give you lots of insight into what is actually going on behind the scenes when your application is running. Instruments also allows you to compare data between different runs of your application. For example, if on your first run of your application you were able to track down inefficiencies in your code, you could fix them and run it again and compare the optimized data to the first. This allows you to see the difference over time as you optimize your application.

iPhone Configuration Utility

The iPhone Configuration Utility is an application available for both Windows and OS X and can be downloaded for free from the Apple Web site. It is a great application to manage profiles and configurations of your device. When your device is connected, you can add and remove provisioning profiles, install and uninstall applications, and view console messages. The iPhone Configuration Utility is great for configuring multiple devices with the same settings. Because it is designed to help enterprise users, you are able to create configuration profiles that contain WiFi settings, VPN configurations, email account settings, a calendar, and certificates that allow your device to communicate with enterprise systems. To download the iPhone Configuration Utility, go to http://support.apple.com/kb/DL926 for Windows or http://support.apple.com/kb/DL851 for Mac OS X.

iTunes

iTunes is the most widely used digital music application. This book discusses two features of the iTunes application. The first has to do with managing your device. You can use iTunes just like many of the other applications mentioned earlier for adding and removing applications and provisioning profiles. The other feature covered here is the iTunes App Store. This is where your applications will be available for download and purchase by other users. Later in the book you will find out about everything that is necessary to successfully submit your application.

iPhone Dev Center

The iPhone Developer Program Portal is the main online area for your iPhone development account. From here you can download the latest iPhone SDK, developer tools, and any beta versions of iOS. There is also an enormous amount of sample code that can be downloaded. Unfortunately, they are all in Objective-C, but some of the samples can be applied and converted to ActionScript 3. There are also a lot of user guides that talk about the different tools and about how different features of the device work. I encourage you to go through as much of the documentation as you can to get a better understanding of the anatomy of an iPhone application.

iPhone Developer Program Portal

You can access the iPhone Developer Program Portal when you log into your iPhone developer account. The Developer Program Portal takes you through the necessary steps to get your applications onto your device so that you can test and distribute them. The Program Portal allows you to add and invite new team members and create certificates and provisioning files. You can also add devices to your account so that you can install your applications on them.

iTunes Connect

iTunes Connect, which can be accessed from the Developer Program Portal, provides you with a set of tools to manage your applications in the App Store. When your application is completed, you can submit it to Apple through iTunes Connect. Furthermore, you can set up any needed banking and tax information if you plan on selling your applications. After your application has been approved by Apple and is ready for sale, iTunes Connect provides you with a suite of Sales and Trend reports that will show you where and how often your application is being downloaded.

Introducing the Available APIs

With the ability to publish iPhone applications from Flash CS5 comes a set of new APIs that enable you to take advantage of some of the features the iPhone has to offer. However, Adobe's strategy is not to support only the iPhone platform but as many platforms as possible. This is the reason you may not see as many iPhone-specific features as you might like or think. Adobe is being very pragmatic about what new features it introduces and how its APIs will look on future platforms, mobile or otherwise. Adobe's goal is to provide one consistent API for all platforms. For example, the ActionScript code should be the same for accessing a camera whether you are developing applications for the Web, desktop, iPhone, or any other future supported platform such as Android.

Accelerometer

The new `Accelerometer` class, which can be found in the `flash.sensors` package, gives you the ability to interact with the accelerometer that is built into the device. The iPhone accelerometer is a three-axis accelerometer capable of measuring both acceleration and gravity. iOS uses the accelerometer to detect its rotation as well as any movements such as shakes.

GeoLocation

The `flash.sensors.Geolocation` class enables you to interact with the device's location sensor. Using this class allows you to retrieve the location of your device anywhere in the world. Coordinates are reported to you in the form of latitude and longitude. There are differences in the ways each device figures out your location. The iPhone 3G and 3GS have actual GPS chips onboard, which provide the most accurate readings. However, the iPod touch does not have a GPS onboard but uses WiFi positioning services in order to figure out your location. It is important to understand that every device that has iOS installed does not necessarily have the same location sensor and that accuracy will differ greatly between devices.

Camera Roll

The `flash.media.CameraRoll` class enables you to save a `BitmapData` instance to the iPhone's camera roll. This feature is currently only supported on the iPhone, and at the time of writing this, there is no fully supported way to load any of the images into your application.

Stage Orientation

There are a few new classes and methods to help handle stage orientation changes and updates. The `flash.display.StageOrientation` class defines a set of valid orientations in which the stage can be set. The `flash.display.Stage.setOrienation` method, which is new in AIR 2.0, allows you to set the orientation of the stage based on one of the static properties in the `StageOrientation` class. And finally there is a `flash.events.StageOrienationEvent` class, which allows you to listen for when the stage orientation is changing and has changed.

Touch Event

The `flash.events.TouchEvent` class is used to detect when a user touches the screen with his or her finger. The `TouchEvent` class is very much the alternative to the `MouseEvent` class but for touches. As well as being available on the iPhone, the `TouchEvent` class is also available on AIR applications in Windows 7 with a touch-enabled screen.

NetConnection

Adobe has added `NetConnection` support for iPhone applications. The `NetConnection` class gives you the ability to connect to a Flash Media Server to create peer-to-peer applications, as well as view streaming video on your device. It will also allow you to communicate with Flash Remoting and the AMF protocol.

Gesture Transform Event

The `flash.events.GestureTransformEvent` class is used to detect specific user interactions with multiple fingers. Touch-enabled interfaces are still very much in their infancy; however, there is already a standard set of gestures that a user understands and expects when interacting with your applications. The `GestureTransformEvent` class can detect four different types of gestures: swipe, rotate, pinch and zoom, and pan. Swipe detects a single finger swiping across the screen in the left, right, up, or down direction. The rotate gesture allows you to place two fingers on an object and rotate one finger around the other to rotate the object. The pinch and zoom gesture enables you to zoom in and out of objects by moving your fingers closer or farther apart from each other. Finally, the pan gesture lets you pan an object in any direction with two fingers. Some of these gestures are supported in Windows 7, OS X 10.5.3, and Windows Mobile 6.5, as well as on the iPhone. If you plan on supporting more than one platform, you should double check which gestures are fully supported on each before starting development.

FLV

FLV is Adobe's Flash Video format. There are two methods of playing an FLV video on the iPhone. The first is to import the video onto the Timeline in an FLA file. This will create a `MovieClip` with your video in it, and you will be able to control it just as you would a normal `MovieClip`. The second method is to bundle the file with the application and load it at runtime. Both of these methods are covered in Chapter 8, "Working with Video," later in the book.

SharedObjects

`SharedObject`s are Flash's version of a browser cookie. They allow your Flash application to save user data on the device. This allows your application to load the save data when you come back to the application.

SQLite Database

The SQLite database is the most widely deployed SQL database in the world. When Adobe released AIR 1.0, it included the ability to communicate with SQLite databases from your applications. This gives you the ability to save large and complex data locally on the device. The iPhone also has SQLite libraries to develop with. A lot of the data on your phone is stored in SQLite databases, such as your Contacts and Call History. Adobe has given us the ability to create, save, and load data from a SQLite database.

Check What APIs Are Not Available

Apple has placed certain restrictions on what developers can and cannot do when creating their applications. This allows Apple to have control over the quality and stability of applications that are available in the App Store. Some of these restrictions have forced Adobe to restrict access to certain ActionScript 3.0 APIs when developers are creating iPhone applications.

Pixel Bender and HTML Loader

In the App Store terms of use, it states that applications cannot run interpreted code. This is the reason that Adobe had to compile a Flash project to native ARM assembly instead of having your SWFs running on an iPhone version of the Flash Player. This is the same reason that Pixel Bender and `HTMLLoader` are not available. Currently, the only way to get Pixel Bender kernels to work on the iPhone is to include the code from the Flash Player, which interprets the kernel. `HTMLLoader` is the same concept. You may be thinking that iPhone applications can already display HTML pages. This is true; however, Adobe has yet to implement the functionality for providing developers access to it. If Adobe were able to bundle its existing code used to display HTML pages, it probably would have, but Apple's terms of use makes everything a little trickier.

Camera

One of the main features of the iPhone is its onboard camera. You can take pictures — and video on the 3GS — and save them to your camera roll. Currently, there is not access to the camera in order to take pictures. However, there is the ability to take a snapshot of the screen and save it to the camera roll. The final piece of functionality currently not fully supported is being able to select an image from the camera roll. This is by far the highest requested feature by developers, and Adobe will be revisiting this feature in the future.

New AIR 2.0 Networking Classes

With the most recent release of AIR 2.0, Adobe has introduced some new networking classes. The `ServerSocket` class enables you to create your own socket server and have other AIR clients connect to it via the `Socket` class. This feature allows you to pass data to and from applications. The `ServerSocket` class is currently not available, but you can create a `Socket` in your iPhone application and connect to one running on your computer.

The `DatagramSocket` class enables you to send and receive UDP data. This can be used to create peer-to-peer applications or gaming. It is more unreliable than a TCP socket because you cannot guarantee the order of the data you will receive and lost packets are not retransmitted or even detected.

Adobe has also introduced a set of classes that can be used in conjunction with the new socket and server classes. For example, the `NetworkInfo` class provides you with a list of all available network interfaces available on the current machine. From those you are able to find out what type of interface they are as well as the IP for the address. This comes in handy when you are creating peer-to-peer applications where you might not know the IP address of the other peer.

There is technically no reason why these classes are not part of the Flash-to-iPhone offering, and they have been a highly requested feature. Adobe does plan on implementing these sometime in the future — when, however, is very hard to say.

Microphone

Support for the microphone is also missing in the iPhone Packager. AIR 2.0 introduced the ability to receive the sample data from the microphone, which would enable you to encode the data and save it out. This would allow developers to create voice-recording applications or allow you to use voice as an additional way to interact with your device.

Push Notifications

Push notifications enable you as a developer to send notifications to users who have downloaded or purchased your applications. The great thing about this is that you can receive notifications even when you are not running the application. For example, an instant messaging application could notify you when someone from your contact list has come online. Push notifications are currently unavailable for all Flash CS5 iPhone applications. This does remain one of the most requested features, and Adobe is looking at providing it in an update to the iPhone Packager after the Flash CS5 launch.

iPhone Standard Controls

One of the great things about iPhone applications, and even OS X desktop applications, is the design and user experience. Apple spends a lot of time creating controls and components for developers to use so that applications have a consistent look and feel, no matter the developer. Currently, there is not an iPhone-specific component set, nor do you have the ability to use the ones in the iPhone SDK. However, there are some PSD templates online that people have provided that have most of the controls. This is not an ideal solution because you will have to still develop the controls, but the more you can make your app look like an application built with the iPhone SDK, the better.

AS1 and AS2

This may come as a surprise to some developers, but there are still many developers who have not adopted AS3 and are still using AS2 or AS1. If you are one of these developers or have some older projects that you are looking to convert, you will need to start learning AS3 and start converting your projects. Currently, the iPhone Packager works only with AS3, and there will not be support for any earlier version of ActionScript. All the tasks and code in this book are in ActionScript 3, so if you are not familiar with it, I suggest you read up on ActionScript in order to get up to speed first; I recommend the book *ActionScript: Your visual blueprint for creating interactive projects in Flash CS4 Professional,* available from Wiley Publishing.

In App Purchase

One thing that has really helped developers sell applications is the availability of a free version of your application. These are sometimes referred to as *Lite versions.* However, you were not able to really tell if any of the customers who played the Lite version actually bought the full version of the application. To help with this, Apple introduced In App Purchase. This allows you to set up a mini App Store inside your game. Now you can provide a Lite version of your application and have the customer purchase new levels or items from within the game. The same App Store rules apply and Apple takes 30% of any transaction. However, this allows you to keep your users engaged in your application and offers you control over what and how they upgrade. Currently, In App Purchases are not available for Flash CS5 iPhone applications. Adobe understands that this is a key feature that will allow users to monetize their applications. Adobe is currently looking at ways that it can provide this feature in the iPhone Packager in future updates.

Become an iPhone Developer

Before you start with any of the topics in this book, you will want to enroll in the Apple iPhone Developer Program. There are three different programs you can enroll in. The free version allows you to download the iPhone SDK and develop applications, but it does not allow you to install applications to your device. The standard program, which is $99, allows you to develop applications, install them on devices, and distribute them in the App Store. The enterprise program, which is $299, is for companies who are looking to develop and distribute in-house applications. Because this book demonstrates developing applications in Flash, you will need to install your applications on the device in order to test them. Some features support debugging on the desktop, but others such as the accelerometer and multitouch will need to be installed on the device in order to be tested.

Chances are that the $99 standard program will be exactly what you need. The standard program comes in two forms, Individual and Company. If you choose Individual, you will not be able to add any team members to your account later on. After you have signed up and enrolled in a program, Apple will send you an activation email that outlines everything you need to do to get your account up and running. If you plan on selling your application in the App Store, you will need to set up your banking information with Apple so that you can get your share of any apps you sell. However, you will not have to set up your banking information if you plan to distribute only free applications.

It is important to know that there is an approval process for enrolling in a developer program, and it is best not to wait until your application is finished to sign up.

Become an iPhone Developer

Register As an Apple Developer

1. In a browser, go to http://developer.apple.com/programs/start/register/create.php.

2. Click Create an Apple ID.

3. Click Continue.

   The Personal Profile page appears.

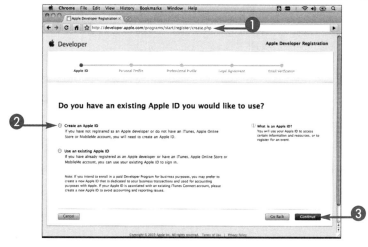

4. Complete the Personal Profile form.

5. Click Continue.

   The Professional Profile page appears.

6. Answer all the professional profile questions.

7. Click Continue.

   The Legal Agreement page appears.

8. Click here to check the box.

9. Click I Agree.

   The Email Verification page appears.

   An email is sent to you with a verification code.

10. Enter your verification code.

11. Click Continue.

    The Success page appears.

12. Click Continue.

    The Member Center page appears.

Enroll in the iPhone Program

⑬ Go to http://developer.apple.com/programs/start/standard/create.php.

⑭ Click I'm Registered As a Developer with Apple.

⑮ Scroll down and click Continue.

The Enrollment Selection page appears.

⑯ Click your enrollment type, such as Individual.

Note: You may be asked to log into your account.

⑰ Complete any account profile questions.

The Billing Information page appears.

⑱ Enter your billing information.

⑲ Click Continue.

The Select Your Program page appears.

⑳ Click iPhone Developer Program.

㉑ Click Continue.

The Review & Submit page appears.

㉒ Review your information.

㉓ Click Continue.

The Program License page appears.

㉔ Accept the terms.

㉕ Click I Agree.

The Purchase Page appears.

㉖ Click Add to Cart.

An activation email will be sent to you.

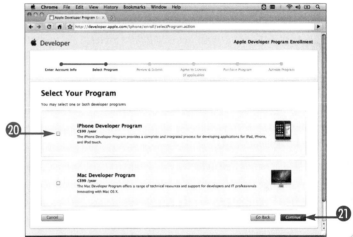

Extra

After you have signed up as an iPhone developer, you will want to install the iPhone SDK. Unfortunately, this is only available for Intel-based OS X machines, with Leopard or later installed. Installing the SDK is not a requirement for creating applications with Flash CS5 and is completely optional. However, it does provide you with some great debugging tools that are covered later in this book. There is also a great deal of information about the iPhone platform in the Developer Program Portal. There are many articles that explain how the iPhone works with lots of details about the hardware itself. There are also several code examples that can be ported to Flash if the feature is available. Even if you never intend to write your applications with the iPhone SDK provided by Apple, I strongly recommend reading through all the "Getting Started" articles in the Developer Program Portal. They will give you a better understanding of what makes a great iPhone application and the types of things Apple is looking for when it reviews your application.

Generate a Certificate Signing Request on Mac OS X

Every application needs to be signed by a valid certificate before it can be installed on an iPhone or iPod. You can request development certificates from the iPhone Developer Program Portal. A development certificate is used only during the development process and is valid only for a specified period of time. Apple can also revoke the certificate before it expires.

You will first need to generate a certificate signing request (CSR) using the Certificate Assistant in the Keychain Access application that comes with Mac OS X. The Keychain Access application can be found in the /Applications/Utilities directory on your hard drive.

Make sure that Online Certificate Status Protocol and Certificate Revocation List are both set to Off in Keychain Access's Preferences window, under the Certificates tab.

When you create your request, the Certificate Assistant will ask you to enter your email address and a common name. When entering this information, make sure to use the same email address and name that you used to sign up for your iPhone developer account. The CA Email Address field is not required and should be left blank.

If the Let Me Specify Key Pair Information check box is present, be sure to check it so that you can select the appropriate information. If for some reason it is not there, you should still be able to continue without issue.

This process simultaneously creates a public and private key that will establish your iPhone developer identity. The Certificate Assistant creates a CSR file, which you will upload through the Program Portal in order to request your certificate.

Generate a Certificate Signing Request on Mac OS X

1. Open the Keychain Access application by double-clicking its icon in the /Applications/Utilities folder.

2. Click Keychain Access → Preferences.

   The Preferences dialog box appears.

3. Click the Certificates tab.

4. Click here and select Off.

5. Click here and select Off.

6. Click here to close the dialog box.

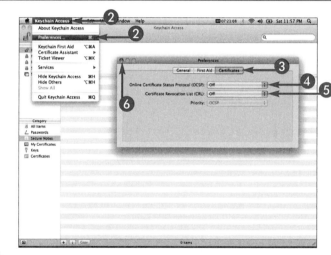

7. Click Keychain Access.

8. Click Certificate Assistant.

9. Click Request a Certificate From a Certificate Authority.

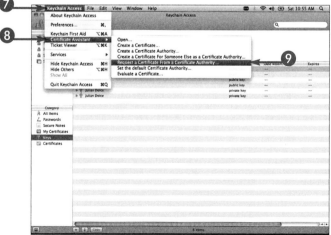

12

The Certificate Assistant appears.

⑩ Type your email address.

⑪ Type your name.

⑫ Click Saved to Disk.

⑬ Click Let Me Specify Key Pair Information.

⑭ Click Continue.

A dialog box appears.

⑮ Select the location to save your certificate signing request.

⑯ Click Save.

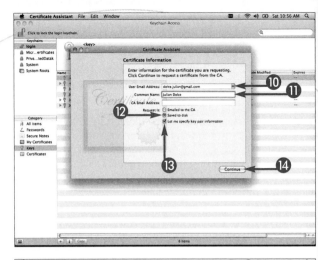

The Key Pair Information page of the assistant appears.

⑰ Click here and select 2048 Bits.

⑱ Click here and select RSA.

⑲ Click Continue.

Your certificate signing request is created.

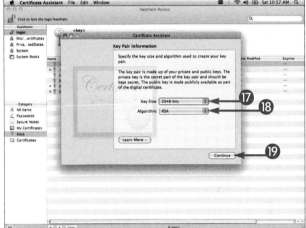

Extra

A certificate signing request is like an application form for a certificate. When you create your certificate signing request, the Keychain Access application creates a public and private key. The request contains the information that you entered, such as your name and email address, as well as your public key. Your private key is not included in the request but is used to digitally sign the entire request.

If you have done any Web site administration before, you may have some experience with creating a certificate signing request. The same process is used when applying for an SSL certificate so that your Web server can communicate over a secure socket layer (SSL). SSL technology protects Web sites and allows visitors to trust that their data is properly encrypted.

Processing credit card information for an online shopping site, logging in, and submitting personal information such as drivers' license numbers or banking information are all examples of when you would use or interact with SSL.

In most modern Web browsers, a lock icon will appear in or near the address bar, signaling that the site is using a secure certificate. Clicking this lock will give you more information about the certificate and what type of encryption it is using.

Generate a Certificate Signing Request in Windows

Every application needs to be signed by a valid certificate before it can be installed on an iPhone or iPod. You can request development certificates from the iPhone Developer Program Portal. A development certificate is used only during the development process and is valid only for a specified period of time. Apple can also revoke the certificate before it expires.

You will first need to generate a certificate signing request (CSR) file using OpenSSL. You can download the application from the OpenSSL Web site, www.openssl.org/related/binaries.html. After you have installed the application, you will use the command line to create your private key and CSR file. Be sure not to ignore any error messages as they may still produce files that are unusable. If you see any error messages, check your syntax and try the command again.

One common error that you may see is "Unable to write 'random state.'" This error suggests that there were not enough privileges to write a file to a certain directory. This is usually caused by the command prompt application not being run by the administrator.

After your private key has been created, you can create your .csr file. When you creating your request, you will be asked to enter your email address and a common name. When entering this information, make sure to use the same email address and name that you used to sign up for your iPhone developer account.

After your certificate signing request file has been created without any errors, you can upload it to the iPhone Developer Program Portal.

Generate a Certificate Signing Request in Windows

1. In an Explorer window, navigate to your Windows System32 directory.
2. Right-click the cmd.exe file.
3. Click Run As Administrator.

The Administrator window appears.

4. Navigate to the bin directory of your OpenSSL install, such as cd C:/OpenSSL/bin.
5. Type **openssl genrsa –out**.
6. Type the name for your key, such as mykey.key.
7. Type **2048**.
8. Press the Enter key.

   Your private key is created.

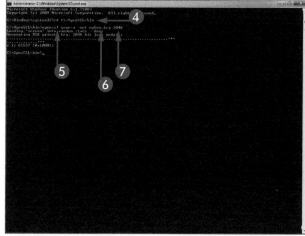

14

9. Type **openssl req –new –key**.

10. Type the name of your .key file, such as mykey.key.

11. Type **–out CertificateSigningRequest. certSigningRequest –subj** "/ **emailAddress=**.

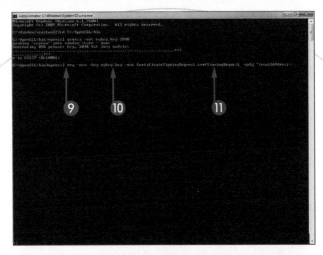

12. Type your email address followed by a comma.

13. Type **CN=**, your name, such as Julian Dolce, and then a comma.

14. Type **C=** and your country code, such as US.

15. Close the quotes and press the Enter key.

Your certificate signing request is created.

Extra

A certificate signing request is like an application form for a certificate. The request contains the information that you entered, such as your name and email address, as well as your public key. In the steps here, you created a private key file, which is not included in the request but is used to digitally sign the entire request.

If you have done any Web site administration before, you may have some experience with creating a certificate signing request. The same process is used when applying for an SSL certificate so that your Web server can communicate over a secure socket layer (SSL). SSL technology protects Web sites and allows visitors to trust that their data is properly encrypted.

Processing credit card information for an online shopping site, logging in, and submitting personal information such as drivers' license numbers or banking information are all examples of when you would use or interact with SSL.

In most modern Web browsers, a lock icon will appear in or near the address bar, signaling that the site is using a secure certificate. Clicking this lock will give you more information about the certificate and what type of encryption it is using.

Upload Your Certificate Signing Request

After you have created a certificate signing request, in the form of a .csr file, you are able to send it to Apple. If you have not created a certificate signing request file, see the earlier section "Generate a Certificate Signing Request on Mac OS X" or "Generate a Certificate Signing Request in Windows," depending on your operating system type.

You can upload your request through the iPhone Developer Program Portal in the Certificates section. Make sure that you are on the Development tab before clicking the Request Certificate button. You can now select the .certSigningRequest file that you created when creating your certificate signing request and upload it to Apple.

After Apple has verified your CSR, it will create a digital certificate file from the information that was included in your request. However, before you can download it, your request will go to your team agent and any team admins who have the authority and responsibility to approve and reject your request. Chances are, if you signed up as an individual, you are the team agent and can approve your own certificate. However, if you are part of a bigger team, you may have to wait until your certificate is approved by one of them in order for you to download it. For a better understanding of the roles and responsibilities of the different types of team members, download the Program Portal User Guide on the home page of the Program Portal. Your team agent and team admins will receive an email notifying them that they need to approve your certificate. After it is approved, you will be sent an email notifying you that it is ready to be downloaded from the Program Portal.

Upload Your Certificate Signing Request

① In a Web browser, go to the iPhone Developer Program Portal Web site, http://developer.apple.com/iphone/manage/overview/index.action.

② Login into the portal.

The Welcome page appears.

③ Click Certificates.

The Current Development Certificates page appears.

④ Click Request Certificate.

The Create iPhone Development Certificate page appears.

5. Click Choose File.

   A dialog box appears.

6. Browse to the CSR file that you created in section "Generate a Certificate Signing Request on Mac OS X" or "Generate a Certificate Signing Request in Windows."

7. Click OK.

   You are returned to the Create iPhone Development Certificate page.

8. Click Submit.

   Your certificate signing request is uploaded to Apple.

9. To download your certificate, click Download on the Current Development Certificates page.

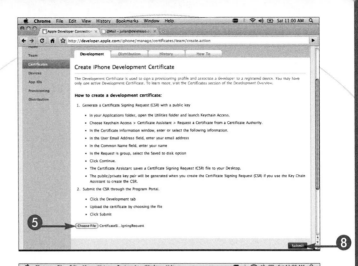

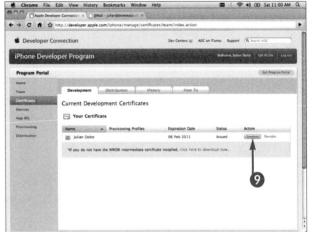

Extra

The iPhone Developer Program Portal allows you to administer the development of your applications. It contains the ability to add and remove team members and request and create all the necessary files that you need in order to install your application on your device. It is also host to a wealth of information that will help you get started in creating applications. I highly encourage you to watch the videos and read the users' guides that can be accessed from the Program Portal home page. You can also request support from Apple. With each developer account, you receive two engineering consultations per membership year. Also, there is a fairly extensive frequently asked questions area on various topics that pertain to the iPhone Developer Program. If you ever have any questions about a topic, be sure to check these, as there is a good chance that you will find your answer there.

There is also an App Store resources section that has information on helping you promote your application, becoming an iTunes affiliate so that you can earn commission on your sales, and app sales and trends.

Create a P12 Certificate on Mac OS X

After you have downloaded your certificate from the iPhone Developer Program Portal, you can convert it to a P12 certificate. This is the file that you will add to your Flash files in order to digitally sign them. If you have not created a valid iPhone developer certificate yet, see the earlier section in this chapter "Generate a Certificate Signing Request on Mac OS X" or "Generate a Certificate Signing Request in Windows" for more details. Find the instructions for your operating system, as there are different instructions for Windows and Mac OS X.

For Mac OS X, you use the Keychain Access application to convert your certificate to a P12 one. Before conversion, you can find your CSR in the login items under the Keychains category. In the listing of keys, you will see the public key and private keys that you created earlier, during the certificate signing request creation process. If you expand your private key, you will see the iPhone developer certificate that you downloaded from the Program Portal.

Selecting the certificate will give you more information about it, particularly when it expires. This information is important, and you will need to go through the process of creating a new one when it expires.

As you are creating the P12 certificate, you will be prompted to create a password to associate with your certificate. Make sure that the password is something that you will remember because you will need it in order to have Flash digitally sign your applications.

Create a P12 Certificate on Mac OS X

1 In a Finder window, double-click the certificate file that you downloaded from Apple.

Note: *See the preceding section, "Upload Your Certificate Signing Request," for more information.*

The Keychain Access application opens, and your certificate is installed in it.

2 Click login.

3 Click Keys in the Category list.

4 Click here to expand your private key.

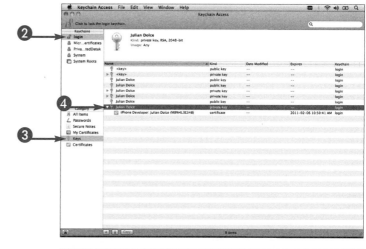

5 Right-click the private key that you installed.

6 Click Export *"Key Name,"* in which *Key Name* is the name of your key.

The Save As dialog box appears.

7 Click here and select Personal Information Exchange (.p12) for the file format.

8 Click Save.

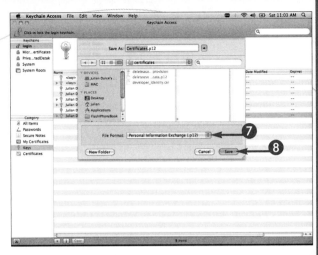

A password dialog box appears.

9 Type in a password.

10 Confirm your password.

11 Click OK.

Your certificate is converted to a P12 one.

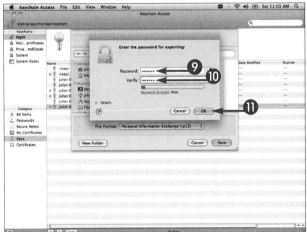

Extra

Make sure to save all the files that you created in this section in a safe place. They will come in handy in the future if you need to reinstall your operating system at any point. Because the files are not that big in size, you can even email them to yourself in case your hard drive fails. However, if you ever have a hardware failure and lose your certificate, you can download it again from the Program Portal. Also, if you have multiple computers that you develop on, you can use the same files on all of them in order to sign your applications. It is important to note that when it comes time to distribute your application for testing or submitting, you will need to create a distribution certificate. For more details on how to do this, see Chapter 18, "Deploying Your Application."

Create a P12 Certificate in Windows

After you have downloaded your certificate from the iPhone Developer Program Portal, you can convert it to a P12 certificate. This is the file that you will add to your Flash files in order to digitally sign them. If you have not created a valid iPhone developer certificate yet, see the earlier section in this chapter "Generate a Certificate Signing Request on Mac OS X" or "Generate a Certificate Signing Request in Windows" for more details. Find the instructions for your operating system, as there are different instructions for Windows and Mac OS X.

Just as you used OpenSSL to create your certificate signing request, you will use it to create a personal information exchange file (.p12). This is the file that you will add to your Flash files in order to have Flash digitally sign them.

Before you start, make sure to place your development certificate that you downloaded from the Program Portal to the OpenSSL bin directory. This is not necessary in order to complete the following tasks, but it will make it easier to reference your files.

The first step is to create a PEM file from your Apple certificate. PEM is an X.509 Base64 encoded distinguished encoding rules (DER) certificate. X.509 is a public key infrastructure standard.

From your PEM file and your private key file, you can use OpenSSL to create your .p12 file. Upon successfully creating the .p12 file, you will be prompted to enter a password. This password can be anything you like; however, make sure that it is something you will remember. You will be required to enter this password when you want to have the iPhone Packager publish your iPhone applications.

Create a P12 Certificate in Windows

① In Explorer, navigate to your Windows System32 directory.

② Right-click the cmd.exe file.

③ Click Run As Administrator.

The Administrator window appears.

④ Navigate to the bin directory of your OpenSSL install, such as cd C:/OpenSSL/bin.

⑤ Type **openssl x509 –in**.

⑥ Enter the path to the .cer file you downloaded, such as developer_identity.cer.

⑦ Type **–inform DER –out**.

⑧ Enter an output filename, such as developer_identity.pem.

⑨ Type **–outform PEM**.

⑩ Press Enter.

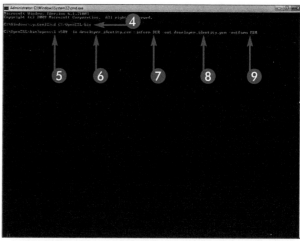

⓫ Type **openssl pkcs12 –export –inkey**.

⓬ Enter the path to your key file, such as **mykey.key**.

⓭ Type **–in** and enter the path to your .pem file, such as developer_identity.pem.

⓮ Type **–out**.

⓯ Enter a name for the .p12 that will be created, such as iphone_dev.p12.

⓰ Press the Enter key.

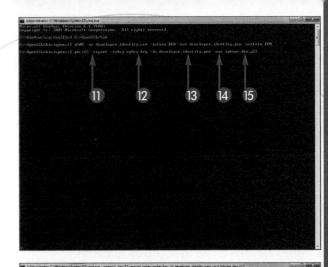

⓱ Create a password.

⓲ Confirm your password.

Your certificate is converted to a P12 one.

Extra

Make sure to save all the files that you created in this section in a safe place. They will come in handy in the future if you need to reinstall your operating system at any point. Because the files are not that big in size, you can even email them to yourself in case your hard drive fails. However, if you ever have a hardware failure and lose your certificate, you can download it again from the Program Portal. Also, if you have multiple computers that you develop on, you can use the same files on all of them in order to sign your applications. It is important to note that when it comes time to distribute your application for testing or submitting, you will need to create a distribution certificate. For more details on how to do this, see Chapter 18.

Add Devices to Your Account

In order to debug your application on your device, you must add it to your iPhone developer account in the Program Portal. Each device has a 40 character unique device identifier (UDID) string, which is unique to that device. You will need to get your team agent or team admin to enter your device UDID for you if you are not one. Each account can register 100 devices for development purposes a year. Any devices that you later remove still count towards your maximum 100 devices a year.

There are several ways to get the UDID of your device, but the easiest and cross-platform way to retrieve it is through iTunes. You will need iTunes 7.7 or later in order to view the UDID, but chances are you have that if you have iTunes. When you are retrieving the UDID of an iPhone, the identifier information will be visible by default in the iTunes window. However, if you have an iPod, only the serial number of the device is shown by default, but you can click that to find your UDID.

After you have the UDID of your device, you can add it to your account. This is done from the Devices section of the iPhone Developer Program Portal. Make sure to give your devices good identifying names because as time goes on, you will add similar devices to your account, and you will want a quick and easy way to distinguish them from each other.

Add Devices to Your Account

1. Attach your device to your computer and open iTunes.
2. Click your device's name.
3. Click Serial Number to show the UDID number.
4. Press ⌘+C (Ctrl+C) to copy the UDID to the Clipboard.

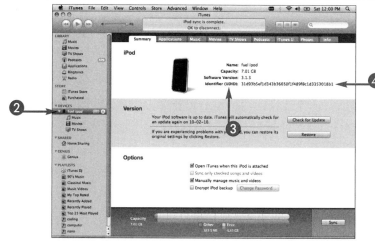

5. In a Web browser, go to the iPhone Developer Program Portal Web site, http://developer.apple.com/iphone/manage/overview/index.action.
6. Log into the portal.
7. Click Devices.

    The Current Registered Devices page appears.

8. Click Add Devices.

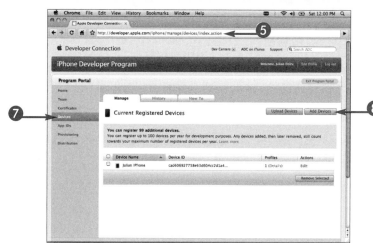

The Add Devices page appears.

9 Type a name for your device, such as fuel iPod 02.

10 Select the Device ID text area and press ⌘+V (Ctrl+V) to paste your UDID.

11 Click Submit.

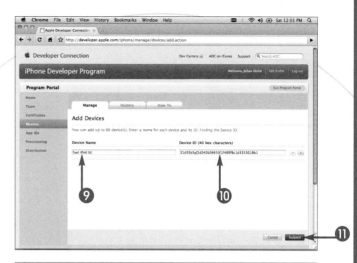

- Your device is now added to your developer account.

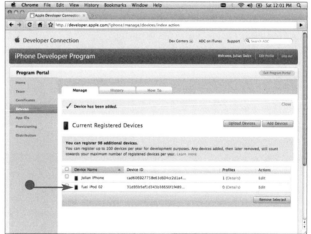

Apply It

If you need to add a lot of devices at the same time, you can do a bulk upload. You can export a selection of devices using the iPhone Configuration Utility application. However, there is a good chance that you will not have the actual devices in order to do this. You can create a .deviceids file in a text file. Here is an example of the format of that file:

```
<?xml version="1.0" encoding="UTF-8"?>
<!DOCTYPE plist PUBLIC "-//Apple//DTD PLIST 1.0//EN" "http://www.apple.com/DTDs/
  PropertyList-1.0.dtd">
<plist version="1.0">
<dict>
                <key>Device UDIDs</key>
                <array>
                    <dict>
                        <key>deviceIdentifier</key>
                        <string>UDID GOES HERE</string>
                        <key>deviceName</key>
                        <string>iPhone</string>
                    </dict>
</dict>
</plist>
```

Create App IDs

An App ID is a unique identifier that allows your application to communicate with the Apple Push Notification service, store data in the keychain, and communicate with external hardware accessories. At the time of writing, Adobe does not support any of these features for Flash iPhone applications, but your device does not know that, and in order to install your application onto a device, you will need to create a valid App ID.

Each App ID consists of a unique 10-character string prefix generated by Apple, known as the *bundle seed ID*, and a *bundle identifier* suffix, which is entered into the Program Portal. Apple recommends using the reverse-domain naming convention when entering your bundle identifier. You may be familiar with this naming convention as it is a common practice for naming your AS3 packages. A sample App ID may look something like this: 8PZ983M77K.com.deleteaso.mysuperawesomegame.

You can also create wildcard bundle identifiers, which look something like this, 8PZ983M77K.com.deleteaso.*. This is extremely handy when you want to quickly try prototype ideas. For example, if you were not able to do this, you would have to create a new App ID for each task and application in this book. This also allows you to create a suite of applications that share the same keychain access, so you can easily use the same passwords across applications. When using the wildcard method, you simply replace the asterisk with whatever you want when entering your App ID in the iPhone Settings panel in Flash.

Create App IDs

1. In a Web browser, go to the iPhone Developer Program Portal Web site, http://developer.apple.com/iphone/manage/overview/index.action.

2. Log into the portal.

3. Click App IDs.

   The App IDs page appears.

4. Click New App ID.

   The Create App ID page appears.

5. Type a name for your App ID, such as flash iphone development.

6. In the Bundle Seed section, click here and select Generate New.

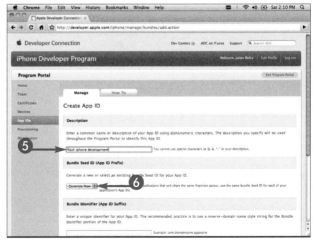

7 Type a bundle identifier, such as com.flashiphonedevelopment.*.

8 Click Submit.

- Your App ID is now created and appears in your App IDs list.

Extra

It is a good idea to start with a wildcard style App ID. This allows you to get going quickly and try out all the new features. After you are ready to start development on an application that you will want to upload to the App Store, log back into the Program Portal and create a full App ID. I always have a wildcard App ID and provisioning profile that is strictly for development. Mine looks something like com.deleteaso.*, which is the reverse domain of my personal development blog. This allows me to create quick samples or tests without having to go to the Program Portal every time I start a new file. For example, during the course of writing the samples for this book, I used the same provisioning profile for every sample. If I made an App ID for every sample, I would have had over 100 provisioning profiles on my device. This would have caused a lot of extra work to manage and maintain them throughout the course of writing this book.

Create a Provisioning Profile

A *provisioning profile* is a file that ties developers and devices to an authorized iPhone development account. A development provisioning profile must be installed on each device that you want to use for testing. A single device can have more than one provisioning profile. Development profiles are valid only for a specified period of time, currently around four months, so you may have to create a new profile if your development goes passed its expiration date.

Provisioning profiles are made up of a set of certificates, devices, and an App ID. You can create a provisioning profile using the iPhone Development Program Portal. Before you can create one, though, you will need to have created your certificate, added at least one device, and created an App ID. If you have not completed all of those steps, see the previous sections of this chapter for more details.

As you add more devices to your account, you can edit your profile to include your new devices. However, you will want to make sure that every device is using the newest provisioning file to avoid any issues when installing the application on the device. After you have created your profile, you will be able to download it so that you can install it on your devices that you selected. Also, you will add it in the iPhone settings of the applications that you develop. For more details on these steps, see Chapter 3, "Developing Your First Application."

Create a Provisioning Profile

1. In a Web browser, go to the iPhone Developer Program Portal Web site, http://developer.apple.com/iphone/manage/overview/index.action.

2. Log into the portal.

3. Click Provisioning.

   The Development Provisioning Profiles page appears.

4. Click New Profile.

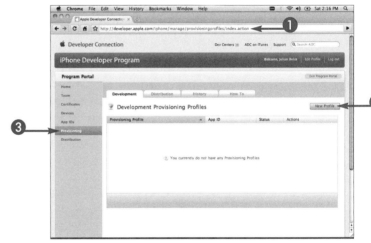

The Create iPhone Development Provisioning Profile page appears.

5. Give your profile a name, such as Flash Development.

6. Select your certificate.

⑦ Click here and select your App ID.

⑧ Select all the devices that you want to include in this profile.

⑨ Click Submit.

● Your provisioning profile now appears in your Development Provisioning Profiles list.

⑩ Click Download to download your provisioning profile.

Extra

It is a good idea to create a general development provisioning profile. To do this, you will want to create a wildcard type App ID and use that as your App ID for your profile. This limits the number of profiles you have to create when you want to quickly prototype an idea or test a simple feature on your device. Only those developers whose Apple device UDID and development certificates are included in the provisioning profile will be able to install and test the application on their device. If you know that you are going to want to test on multiple people's devices, it is a good idea to get their devices and certificates added before you create your development profiles. This will save you from re-creating them later. When you get closer to wanting testers for your application and sending builds to nondevelopers, you can create an ad hoc provisioning file. This will allow users not enrolled in the iPhone Developer Program to test your application and provide you with valuable feedback. You can find more details about how to create ad hoc provisioning files in Chapter 18.

Add Provisioning Files to Your Device

After you have created a provisioning profile, you need to install it onto your device. There are several different ways to do this. The first thing you will need to do is make sure that your device is connected to your computer. The easiest way is to drag the provisioning file into iTunes, and the next time you sync your device, the file will be installed. Alternatively, you can add the provisioning file using the iPhone Configuration Utility. These two methods work both in Windows and on Mac OS X because each application runs on both operating systems. If you are developing on an OS X machine, you can also use Xcode to manage your profiles. In Xcode's Organizer window, you can add profiles by pressing the + button in the Provisioning section of your device and then selecting the .mobileprovision file. This will automatically install the profile on your device, and you will not need to sync it.

When a profile expires, you will probably want to remove it from your device. You can easily do this on the device from the Settings application. However, sometimes you will find that the profile will be re-synced on your device. When the profiles are installed on your device, they are also stored on your machine, so they can be synced when you sync your device with your computer. On OS X, they are located at <user>/Library/MobileDevice/Provisioning Profiles. In Windows 7, they are located at C:\Users\<user>\AppData\Roaming\Apple Computer\MobileDevice\Provisioning Profiles. If you have sync issues, you can go to these folder locations and delete the actual file. This should eliminate any sync issues you may encounter.

Add Provisioning Files to Your Device

1. Open iTunes.
2. Connect your device.

3. Click File.
4. Click Add to Library.

The Add To Library dialog box appears.

5 Click your .mobileprovision file.

6 Click Choose.

7 Click your device's name.

8 Click Sync.

Your device syncs with iTunes.

When your device finishes syncing, the profile will be installed on your device.

Extra

If you installed the provisioning profiles with either Xcode or the iPhone Configuration Utility, you can select the profiles to see some important data. Both applications do an excellent job of alerting you when a profile is about to expire and when it has. Selecting a profile in either application will show a big yellow bar if a profile is about to expire and a big red bar if it has expired. If you are working with a team of developers who all use the same development provisioning profile, it is a good idea to create a new one before it expires. This will save the team lots of headaches scrambling to get a new one created at a critical time in development. Another useful piece of information provided by Xcode and the iPhone Configuration Utility is the profile App Identifier. This tells you what App ID your profile was created with. This saves you from always having to remember or going back to the Program Portal to find it. You can also see which of your devices has the selected profile installed.

Using the Actions Panel

The Flash Actions panel is where all your scripting or ActionScript coding is done within an FLA file. The Actions panel consists of three main sections: the Actions toolbox, at the top left of the panel, which groups similar ActionScript elements together; the Script navigator, at the bottom left, which enables you to jump to your different scripts easily; and the Script pane, on the right, which is where you write your ActionScript code.

The Actions toolbox is a listing of all the internal ActionScript classes and methods provided by Adobe, which allow you to program your Flash application. If there are times when you cannot remember what a class's method name is, you can find it in the list and double-click it to have it added to the Script pane.

You can write scripts on a frame in the Timeline or on an object on the Stage. It is best practice to write only frame scripts because they are easier to find and keep all your code grouped together. There will be times when you will write code on multiple frames and sometimes on different Timelines. The Script Navigator gives you the ability to quickly jump between all the scripts in your file. This can be a big timesaver because it allows you to keep the Actions panel open while writing code.

The Script pane is where you will write all your ActionScript code. It is a basic text editor with some code editor features that will help you be a more efficient programmer. Many of these features can be accessed from the toolbar above the editor window. You can format your code, check its syntax, add and remove comments, and collapse blocks of code.

Using the Actions Panel

Using the Actions Panel to Code

1. Click the frame on the Timeline for which you want to write ActionScript.

2. Click Window.

3. Click Actions.

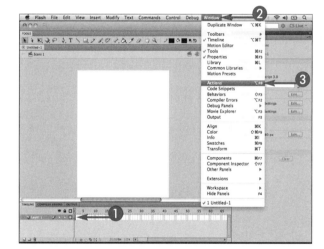

The Actions panel appears.

4. Click an Actions main category, such as Language Elements.

5. Click a subcategory, such as Global Functions.

6. Click an action, such as trace.

   The action appears in the Script pane.

7. Type text in the action, such as "Hello World" for a `trace` action.

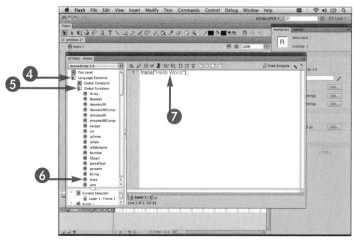

Commenting Code

- Click here to apply block comments to selected code.
- Click here to apply single line comments to a selected line.
- Click here to remove comments.

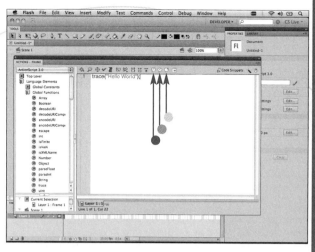

Using Actions Panel Commands

1. Click the additional options button.

   The many commands that you can use while using the Actions panel appear.

Note: Learning the keyboard shortcuts for these commands will help speed up development.

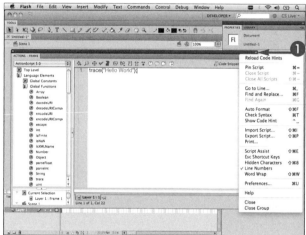

Extra

A new feature introduced in Flash CS5 is the ability to introspect your custom code and provide code hinting. This can drastically speed up development time, as this will save you from typing more than you need to. If you bring up code hinting on an object, it will bring up all its public methods and properties, which acts like an outline for its help file. This saves you from having to remember all of a class's public attributes and gives you the ability to quickly look them up when you need to. To bring up the code hint window as you type, press Ctrl+spacebar. Depending on where your cursor is, different options will appear. If you do not have anything selected, a list of all the available ActionScript classes and objects will appear in the list. If you are on an instance of a class, the list of public methods and properties will appear. When the code hint window appears, you can continue to type what you are looking for, and it will narrow the search for you. You will also notice that when using code completion, it will add any import statements for that class. As you go through the examples in the book, try to use this feature as much as possible, as it is a good habit to pick up.

Create a Skeleton Custom Class

As your scripts become more complex, you will want to create custom classes for them in order to better organize your code. To do this, you will need to create a separate .as file, which you can do in the IDE. However, if you are working on a project with many custom classes, you may want to look at a more full-featured programming environment, such as Flash Builder, to write your ActionScript code.

A custom class consists of a few key parts: The first is a package, which is a way to group similar classes together. If you do not give your class a package name, it resides in the default package of your application. The second part is a constructor, which is the main entry point to your class. The *constructor* is simply a function method with the exact same name as the file or class. When you create a new instance of your class, the code in the constructor will be fired. This is where you will place any initialization that needs to occur for your class to work correctly. It is a good practice to limit the amount of code in your constructors. Having a constructor call a separate `init()` method that does all the initialization is a good way to achieve this. The reason for this is that code in your constructor is not optimized when you compile your application.

After you have the base of a class created, you can add properties and methods to give your class the functionality that you need.

Create a Skeleton Custom Class

① Click File.

② Click New.

The New Document dialog box appears.

③ Click ActionScript 3.0 Class.

④ Click OK.

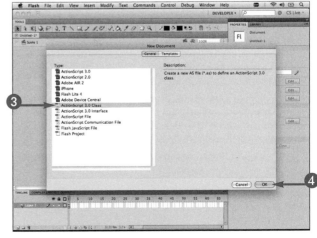

The Create ActionScript 3.0 Class dialog box appears.

5 Give your class a name, such as MyClass.

6 Click OK.

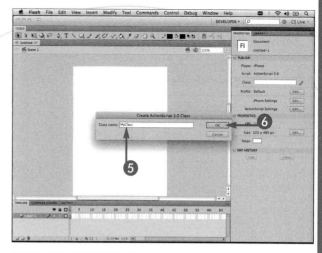

A skeleton class is generated for you.

- You place the package declaration here.
- You place the class declaration here.
- You place the class constructor method here.

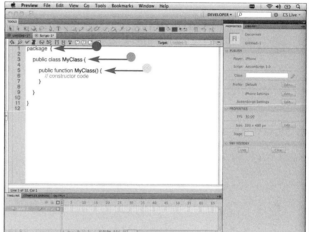

Extra

When naming your packages, it is a good practice to use the reverse-domain naming convention. This makes sure that your classes will not conflict with any other classes created by another developer that have the same name. For example, the URL of my blog is www.deleteaso.com, so my packages would look like this, com.deleteaso. If you give your packages a name, you will also have to place your class file in a set of folders that mimics your package name. In my case, my classes would go in a folder named com/deleteaso. After you have your main package defined, you can create more packages within it to better group similar classes. For example, all of my custom event classes may go in a package named com.deleteaso.events. Try to follow a similar naming convention to group classes to that of which Adobe has adopted for the internal ActionScript classes. This will give other developers an idea of which package to look in for certain functionality.

Set the Source Path

As your projects get more complex and have more custom classes and more .fla files, you are going to want to organize them in a way that makes sense for you. A common way to do this is to have all your classes in a Classes folder and all your .fla files in an src folder. Doing this early in your development will save you lots of time as your project grows.

After you place your .fla files no longer in the same directory as your .as files, you need to tell Flash where to find them so that they can be compiled with your application. To do this, you need to set the source path to the folder where your classes are located.

You can set this individually in each of your .fla files. This is done from the Advanced ActionScript 3.0 Settings dialog box. When the dialog box first appears, the Source Path tab will be selected, and a list of folder locations is shown underneath it. The default location is always set to ., which refers to the same folder in which the .fla file is in. You add your new location to the list. You can also select the folder icon to select the folder in a File Browser window.

You can also set the source path globally for every .fla file that you create. This is done from the ActionScript 3.0 Advanced Settings dialog box accessed from the Preferences dialog box. The process is the same for adding paths globally as it is for an individual .fla file.

Set the Source Path

Set the Source Path for a File

1. Click File.
2. Click ActionScript Settings.

The Advanced ActionScript 3.0 Settings dialog box appears.

3. Click the + button to add a new path.
4. Enter the path to your classes, such as ./Classes.
5. Click OK.

The source path is set for the file.

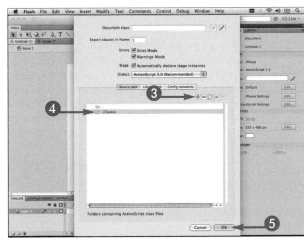

Set the Source Path Globally

1. Press ⌘+U (Ctrl+U).

   The Preferences dialog box appears.

2. Click ActionScript.

   The ActionScript settings appear.

3. Click ActionScript 3.0 Settings.

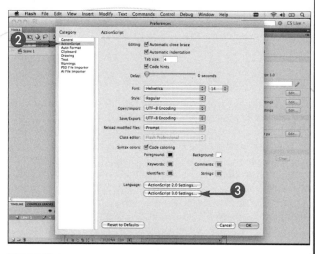

   The ActionScript 3.0 Advanced Settings dialog box appears.

4. Click the + button.
5. Enter the new source path.
6. Click OK.

   You are returned to the Preferences dialog box.

7. Click OK.

   The source path is set globally.

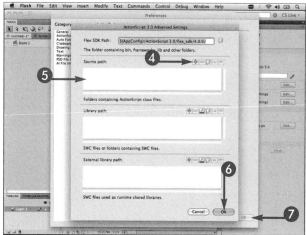

Extra

When setting your source path, it is a good practice to use relative paths instead of absolute paths. This is really important if multiple people are working on the same project on different computers. If you use absolute paths, your class paths will likely not be the same. Using relative paths is easy if you understand a few concepts. If you want to reference a directory that is a level up from the current one, you can use `../`. If you want to reference two levels up, you simply double that, `../../`. So if you wanted to create a relative path for the example shown in this section, you would have used `../Classes`. This tells Flash to go up a level from inside the src folder and select the Classes folder.

Along with a folder for your .fla files and your .as files, it is a good idea to create a folder in which Flash will create all of the .swf files for your project. This folder is usually called bin, or something similar. This will make it easy if you ever have to deploy your .swf files to a server. Make sure to also use relative paths when setting the output location for your .swf.

Create MovieClips

When you want to interact with assets in your Flash project, you will need to create or convert current assets into symbols. Symbols come in a few different forms, the main one being a `MovieClip`. A `MovieClip` symbol consists of a Timeline, which can contain other types of symbols. You can also control the playback of the Timeline independently of other Timelines. `MovieClips` can also be scripted with ActionScript. These are by far the most powerful symbol in Flash, so the majority of the symbols you create will be `MovieClips`.

You can convert any asset that is already on the Stage to a `MovieClip` using the Convert to Symbol dialog box, in which you can name the `MovieClip`, select its type, select the folder in the Library to create it in, and set the registration point of the symbol.

The *registration point* is the location in which your `MovieClip`'s contents will be placed. By default, the registration point is set to the top left, or 0,0. Most of the time this will work; however, there are times when you will want to change this. There are examples later in the book that use the center of the object as its registration point. This is commonly used when you want objects to rotate or scale from its center and not from the top left of the object.

You can also create an empty `MovieClip` by clicking Insert → New Symbol or by pressing ⌘+F8 (Ctrl+F8). After the symbol has been created, Flash will open it in Edit mode so that you can add assets to it.

Create MovieClips

1. Select a shape tool, such as the Rectangle tool.
2. Draw the shape on the Stage.
3. Click the Selection tool.
4. Select the shape on the Stage.

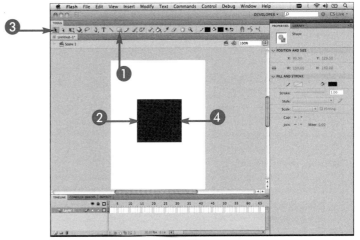

5. Click Modify.
6. Click Convert to Symbol.

Note: You can also use the keyboard shortcut F8.

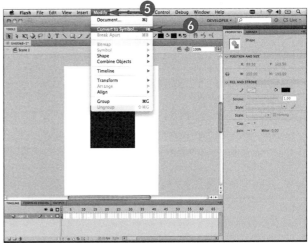

The Convert to Symbol dialog box appears.

7 Give your symbol a name, such as square_mc.

8 Click here and select Movie Clip as the type.

9 Click here and set the registration point of the symbol.

10 Click OK.

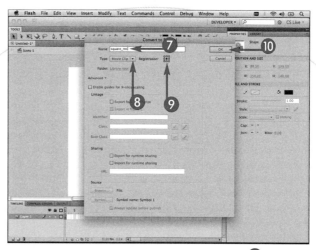

11 Click Window.

12 Click Library to display the Library panel if it is not already checked.

- The symbol you just created appears in the Library.

Note: *You can use the ⌘+L (Ctrl+L) keyboard shortcut to toggle the visibility of the Library panel.*

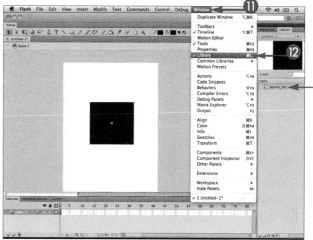

Extra

You can also create a MovieClip with ActionScript. There are many reasons why you would want to create it with code instead of with the Flash authoring environment, which will be come apparent as you work through the other samples in the book. Here is the syntax for creating a MovieClip with code:

```
var mymc:MovieClip = new MovieClip();
```

Another symbol that is similar to a MovieClip is a Sprite. A Sprite is a MovieClip without a Timeline. Oftentimes, if you do not need to do any Timeline animation, you will create a Sprite object instead. These cannot be created in the IDE and must be created with code. Here is the syntax to create a Sprite instance:

```
var mysprite:Sprite = new Sprite();
```

Create Buttons

Whereas `MovieClip`s are great for working with animations and displaying content, buttons are used to allow your users to interact with your application. Button symbols respond to various interactive events, such as rollovers, mouse clicks, and touches when developing touch-enabled applications.

Buttons are simple four-frame interactive `MovieClip`s. The first three frames represent different visual states for the button, and the fourth represents the overall hit area of the button. The play head or Timeline never plays but goes to the appropriate frame when it reacts to the mouse or a touch.

The first frame is the *up state*. This is the frame that is shown when nothing is interacting with it. The second frame is the *over state*. This is the frame that is shown when a user places the mouse cursor over the button. The third frame is the *down state*. This is the frame that is shown when the user presses the mouse button down while over the button. This frame is also shown when a user touches down on the button when developing for touch-enabled devices, such as the iPhone. The fourth and last frame is the overall hit area of the button. This sets the bounding box of the button in which the user can interact with it. This frame is invisible when the file is published.

After your button is created, you can receive events from it when the user interacts with it. This will give you the ability to respond to a user's interaction and update the screen as needed. For more details on handling events, see the section "Work with Events" later in this chapter.

Create Buttons

Create a Button

1. Click Insert.
2. Click New Symbol.

Note: You can also use the ⌘+F8 (Ctrl+F8) keyboard shortcut to create a new symbol.

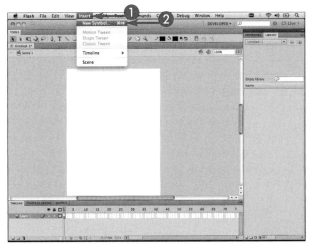

The Create New Symbol dialog box appears.

3. Name your symbol, such as `mybutton_btn`.
4. Click here and select Button as the type.
5. Click OK.

The button is created.

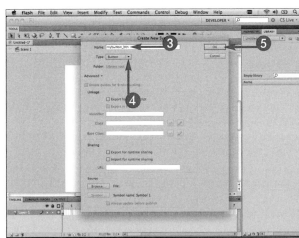

Create the Visual State of the Button

⑥ Select a shape tool, such as the Rectangle tool.

⑦ Select the Up frame in the Timeline.

⑧ Draw a shape.

⑨ Click Scene 1 to exit the button Timeline.

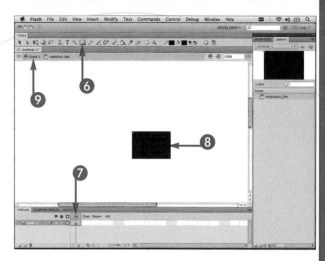

Place an Instance of Your Button

⑩ Click your button in the Library and drag it to the Stage.

An instance of the button is created.

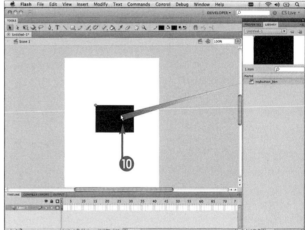

Extra

Button symbols provide you with a very quick and easy way to add interactivity to your projects. However, there will come a time when you will want to use animation or create some more complex button states, instead of the static four frames of a basic button. In order to do this, you may look at creating buttons from `MovieClip` symbols. `MovieClip` and button symbols respond to the same interactions from a mouse or touch. By listening for these events, you can play or show different animations for each state. This is shown in more detail in Chapter 5, "Handling Interaction." If you want to have the mouse cursor turn to the hand when you roll over a `MovieClip` button, simply set the `buttonMode` property to true:

```
mymc.buttonMode = true;
```

Edit Properties in Flash

After you have created some symbols, you can add them to the Stage. When they are on the Stage, you can set many of their properties in the Properties panel.

Setting the instance name of the symbol allows you to reference it through code. Many of the examples later in the book require you to set the instance name for an object. It is always a good idea to name your instances when they are added to the Stage.

You can also adjust the size and position of your object. The `x` property will set your object's position along the x-axis, left to right, and the `y` property will set the its position along the y-axis, top to bottom. Adjusting the `width` and `height` properties will change the size of your object.

There are also many Color Effects settings that you can apply to your object, such as Brightness, Tint, Advanced, and Alpha. The one that you will probably use the most is the Alpha setting. This will set the initial alpha property of the object, or transparency. It is important to note that you can set only one color effect to your object in the Properties panel. However, you can also set them with code, which will allow you to apply multiple effects at once.

You can also apply filters to your object. Flash comes with a bundle of filters, similar to those in Photoshop: Drop Shadow, Blur, Glow, Gradient Glow, Gradient Bevel, and Adjust Color. Unlike the color effects, you are able to apply multiple filters to your objects. You can also save filter presets, which allow you to apply the same set of filters to multiple objects in your file.

Edit Properties in Flash

1. Select a shape tool, such as the Rectangle tool.
2. Draw a shape on the Stage.
3. Click the Selection tool.
4. Select the shape on the Stage.

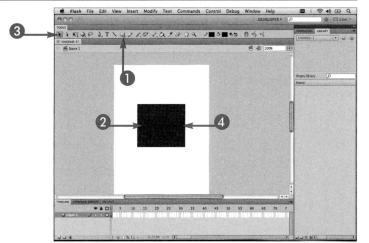

5. Press F8.

   The Convert to Symbol dialog box appears.

6. Give your symbol a name, such as `square_mc`.
7. Click here and select Movie Clip as the type.
8. Set the registration point of the symbol.
9. Click OK.

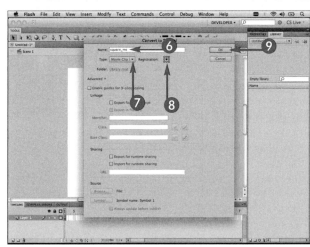

⑩ Click Window.

⑪ Select Properties if the Properties panel is not already visible.

Note: You can also use the ⌘+F3 (Ctrl+F3) keyboard shortcut to toggle the visibility of the Properties panel.

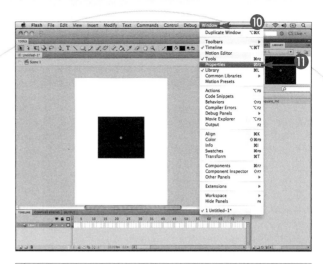

⑫ Give your symbol an instance name, such as square_mc.

⑬ Change the y position.

⑭ Change the width property.

⑮ Click here and select a Color Effect style, such as Alpha.

You can set other properties in the panel as well.

Apply It

As mentioned earlier, the Properties panel is just one way to set your object's properties. You will often want to change these properties at runtime and will need to do so through code. First, make sure that you have given any objects you want to affect an instance name. Here is what setting some of the properties you can set in the IDE would look like in ActionScript:

```
myinstance.x = 100;
myinstance.y = 250;
myinstance.width = 50;
myinstance.height = 64;
myinstance.alpha = 0.5;
```

Add Objects to the Stage with Code

In the Flash authoring environment, adding items to the Stage is simply a matter of dragging items from the Library onto it. This is all well and good, but there will be a time in your project when you want to add objects to the Stage with code. The Stage in this case is called the display list. The *display list* is a hierarchal tree of every visual item that is displayed at any given time. The order of the list is also very important because the lower an item is on the list, the lower the depth that it will appear visually.

In order to add an object to the display list, you will need to create one with ActionScript. You could simply instantiate a new `MovieClip` instance with the following syntax:

`var mc:MovieClip = new MovieClip();`

Alternatively, you can create a new instance of an item that resides in your Library. To do so, you must set the item to Export for ActionScript in the item's Properties panel. When you give it a class name, choose something that well represents it so that it is easy to remember. If you do not already have an .as file for that class, Flash will automatically generate one for you when you compile your project. This will not create an actual file but simply create the class so that it gets compiled with your .swf. To instantiate your Library asset, use the same syntax that you use for a `MovieClip`, except replace `MovieClip` with the class name you gave your Library item.

After you have created an object, you can add it to the display list. You can do this by calling the `addChild()` method on any `DisplayObjectContainer` class. This is the base class for any objects that can serve as display list containers. The two most common are `MovieClip` and `Sprite`.

Add Objects to the Stage with Code

① Create a `MovieClip` symbol and select it in the Library.

Note: *For more details on how to create* `MovieClip` *symbols, see the section "Create MovieClips" earlier in this chapter.*

② Right-click the item.

③ Click Properties.

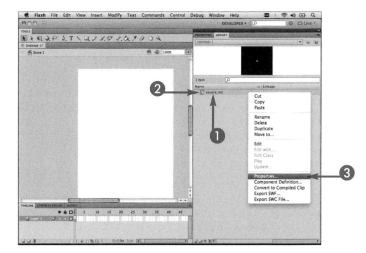

The Symbol Properties dialog box appears.

● You may have to click here to expand the Advanced properties.

④ Click Export for ActionScript.

⑤ Give your symbol a class name, such as `Square`.

⑥ Click OK.

Note: *If a dialog box appears saying, "A definition for this class could not be found in the classpath, so one will be automatically generated in the SWF file upon export," click OK.*

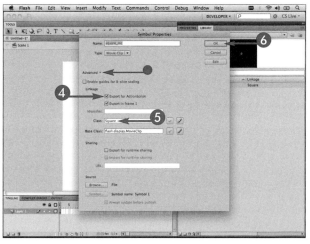

7 Select a frame in the main Timeline.

8 Open the Actions panel.

Note: See the section "Using the Actions Panel" for more information.

9 Create a new Square variable, such as `var mc:Square = new Square();`.

10 Add your object to the Stage, such as `addChild( mc );`.

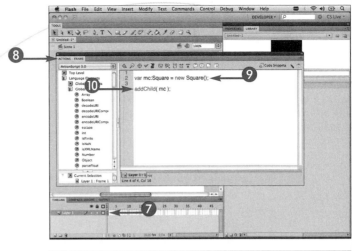

11 Press ⌘+Enter (Ctrl+Enter) to test your movie.

● Your object is now added to the Stage.

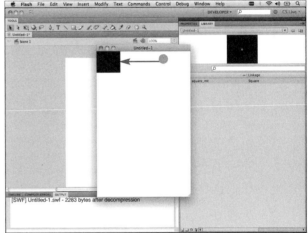

Apply It

When you add a object to the display list using the `addChild()` method, it gets added to the top of the list. This means that the object will be shown on the topmost layer and will overlap any items at lower depths at the same location. There are times when you want to add items at a certain index in the display list. To do this, you can use the `addChildAt()` method. This method is very similar to `addChild()`, with the addition of an extra index parameter. The index parameter specifies which index in the display list you would like to add the object at:

```
var mc:MovieClip = new MovieClip();
addChildAt( mc, 5 );
```

Be careful to make sure that you are adding your object to an index that is in range of the current display list. This means that you cannot add an item to an index that is higher than the total number of children currently on the list. For example, if there are currently four items in the list, you cannot add your item at index 6.

Remove Objects from the Stage with Code

Just as you can add objects to the display list with `addChild()` and `addChildAt()`, you can remove objects with `removeChild()` and `removeChildAt()`. `removeChild()` takes a reference to the object that you want to remove as its only parameter. `removeChildAt()` takes the index of the child in the display list that you want to remove. If you try to remove a child at an index that does not exist, an error will be thrown.

It is important to note that removing an object from the display does not remove it from memory. You can add it back onto the display list later if you like. If you want to make sure that it gets cleared from memory, set the instance of your object to null, after it has been removed. You will also want to remove any other references to your object, such as event listeners. For more details on how to remove event listeners, see the section "Work with Events" later in this chapter.

Being able to re-add your objects back onto the display list is convenient; however, the trade-off is that you will have to be diligent in making sure that it can be cleared from memory. Memory management is not an exact science, and there are many things to take into account. The topic alone could be a whole book in and of itself. There are many great resources and examples online on how to make sure that your objects get cleared from memory or garbage collected. I strongly recommend reading as much information as you can on the subject and finding out what works best for your specific situations.

Remove Objects from the Stage with Code

① Create a `MovieClip` symbol and place it on the Stage.

Note: See the section "Create MovieClips" earlier in this chapter for more information.

② Click the Selection tool.

③ Select the symbol.

④ Give it an instance name, such as `square_mc`.

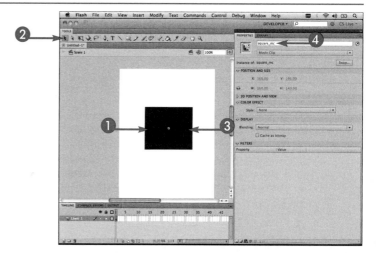

⑤ Click the New Layer button in the Timeline panel.

⑥ Select a frame in the newly created layer.

⑦ Open the Actions panel.

Note: See the section "Using the Actions Panel" for more information.

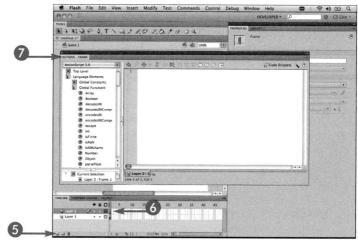

8 Remove the instance from the Stage, such as `removeChild( square_mc );`.

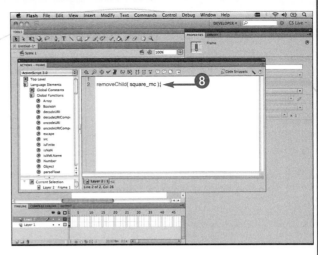

9 Press ⌘+Enter (Ctrl+Enter) to test your movie.

10 Your object is now removed from the Stage.

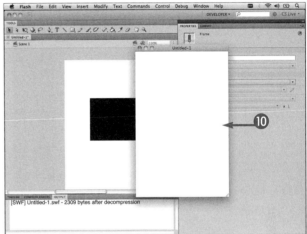

Apply It

The `removeChild()` and `removeChildAt()` methods are great at removing one object from the display list at a time. But what happens if you have hundreds of items that you want to remove at once? You would expect there to be a `removeAll()` method, but this does not exist. If you ever need to write your own, similar method, you can use the following code:

```
while( numChildren ){
    removeChildAt( 0 );
}
```

This removes all the objects from the display list at once.

Work with Events

In ActionScript, an object dispatches an event when an action takes place. Events can be fired when the user interacts with your application, such as clicks a button, or the Flash Player can dispatch them when it is executing specific tasks, such as adding an object to the display list. Events are one of the main mechanics in which one or more objects talk to each other.

You can listen for when events occur by adding an event listener to an object, which dispatches the event. To add an event listener, use the addEventListener() method. The first two parameters, which are required, are the type of event that you want to listen for and a reference to an event handler function.

An *event handler* is simply a function that will be called when the event is dispatched. Every event handler takes one parameter, which is the type of event that it is listening for. The Event object will have all the necessary information that you need in order to respond correctly to an event being dispatched.

Like all things in ActionScript, if you can add events, you can remove them as well. When you no longer need to listen for a particular event, make sure to remove the listener by calling the removeEventListener() method. It is important to remove all unwanted event listeners when you no longer need your objects. This will help make sure that they are cleared from memory by the garbage collector.

Work with Events

① Select a frame in the Timeline in which you want to add your ActionScript code.

② Open the Actions panel.

Note: *See the section "Using the Actions Panel" for more information.*

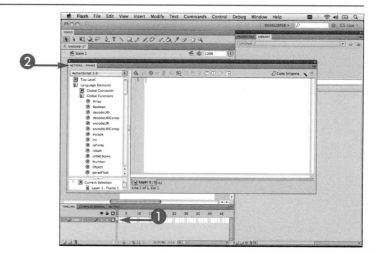

③ Add an event listener to the Stage object, such as stage.addEventListener();.

④ Specify the type of event, such as MouseEvent.CLICK.

⑤ Specify an event handler, such as onStageClick.

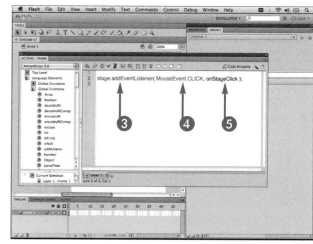

6. Create an event handler method, such as `function onStageClick();`.

7. Add an event object as its parameter, such as `event:MouseEvent`.

8. Add a `trace` statement when clicked, such as `trace( "stage click" );`.

9. Remove the event listener, such as `stage.removeEventListener ( MouseEvent.CLICK, onStageClick );`.

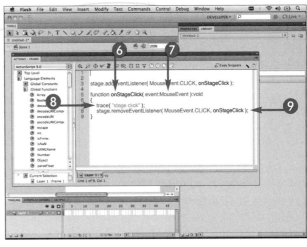

10. Press ⌘+Enter (Ctrl+Enter) to test your movie.

11. Click the Stage of the movie.

● Your `trace` statement appears in the Output panel.

Note: *Clicking the stage a second time will not show the `trace` statement because you removed the listener.*

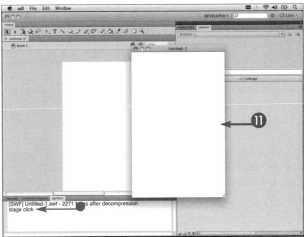

Extra

Now that you know how to add, remove, and listen for events, you can start to dispatch your own events. The `EventDispatcher` class, and any object that is added to the display list, can dispatch events. The `dispatchEvent()` method is what you will use to dispatch your event. It takes an `Event` object as its only argument. This is the same object that will be passed to your event handler that is listening for your event. Here is an example of an event being dispatched when an animation has completed:

```
dispatchEvent( new Event( "animationComplete" ) );
```

As you get more comfortable with events, you can create your own classes that subclass `Event`, which then can be passed into the `dispatchEvent()` method.

Using the Drawing API

The drawing API enables you to draw vector shapes with ActionScript. This is the code equivalent of drawing shapes, such as rectangles and circles, in the Flash authoring environment. You can draw in two different types of objects, Shapes and Sprites. Both of these objects have a graphics property, which is where all the drawing occurs.

The Shape class is a lightweight DisplayObject whose sole purpose is to draw shapes in. The Sprite class can also be used if you need to add items to its display list, whereas the Shape class cannot. However, with this added functionality, the Sprite class will consume more memory.

The Graphics class contains all the methods that you can use to draw vector shapes. There are three steps to drawing a shape: First, you want to set the fill color and alpha of the shape with the beginFill() method. Next, you can use any of the draw shape methods to draw the actual shape. The drawRect() and drawCircle() methods are two of the simpler and more popular shape methods. After you have drawn your shape, close the fill by calling the endFill() method.

The Graphics class also gives you the ability to draw lines with code. First, set the line style that you want to draw with the lineStyle() method. This method allows you to set the thickness, color, and alpha of the line, as well as some other more advanced properties. To draw the line, call the drawLine() method. This draws a line from the current position to the one passed into the method. If you want to move to a new position without drawing a line, you can use the moveTo() method.

When you want to clear your drawing, call the clear() method.

Using the Drawing API

1. Select a frame in the Timeline in which you want to add your ActionScript code.

2. Open the Actions panel.

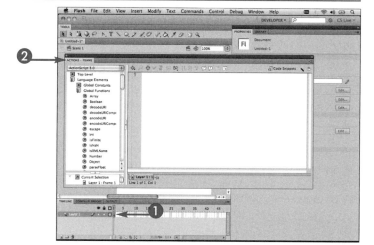

3. Create a Sprite variable, such as `var rect:Sprite = new Sprite();`.

4. Select a fill color, such as `rect.graphics.beginFill( 0x000000 );`.

5. Draw a shape, such as `rect.graphics.drawRect( 0,0,100,100 );`.

6. End the fill, such as `rect.graphics.endFill();`.

7. Add the shape to the Stage, such as `addChild( rect );`.

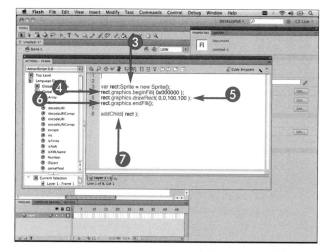

⑧ Create a Sprite variable, such as var line:Sprite = new Sprite();.

⑨ Set the line style, such as line.graphics.lineStyle(5, 0x000000);.

⑩ Set the start position, such as line.graphics.moveTo(0, 200);.

⑪ Draw a line, such as line.graphics.lineTo(100, 200);.

⑫ Add the line to the Stage, such as addChild(line);.

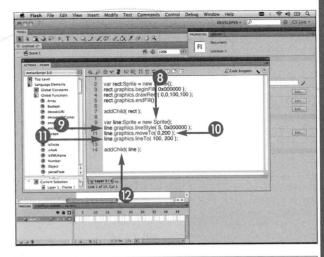

⑬ Press ⌘+Enter (Ctrl+Enter) to test your movie and see your drawings.

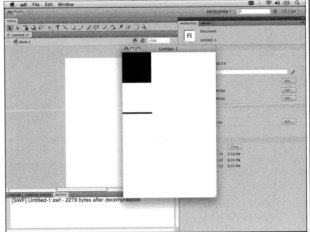

Apply It

When drawing shapes, you can also fill them with a bitmap. For example, say you wanted to create a tile-able background. You could draw a rectangle the size of your objects and fill it with a `BitmapData` object. To create a `BitmapData` object, import an image to the Library, export it for ActionScript, and give it a class name. This will allow you to create a `BitmapData` object of that image. The following example gives the image a class name of `Background`:

```
var bitmapData:Background = new Background();
this.graphics.beginBitmapFill( bitmapData, new Matrix(), true );
this.graphics.drawRect(0,0,stage.stageWidth,stage.stageHeight );
this.graphics.endFill();
```

Using Flash CS5 Help

With the release of CS5, Adobe has created a new AIR application called Adobe Help, which can be launched by pressing F1. The Help application aggregates all the help files for the CS5 products in one central location. With the Help application, you are able to store all your help files locally so that you can access them when you are offline. It also provides an easy way to receive help updates from Adobe when they become available.

There are three sections of the Help application: the Search pane, the Feedback & Rating pane, and the browser pane. The browser pane is a simple Web browser that displays HTML content. There are back and forward buttons that enable you to navigate to previously viewed pages, just as you would in a real Web browser. The URL for the page is also in the top right of the pane, and clicking it will open the page in your default browser.

The Search pane allows you to search your local help content, adobe.com, and many of the community sites. By selecting Community Help in the Search Location drop-down list, you will get results to some of the more popular forum and tutorial sites on the Web. It is like having Google integrated into the Help application.

Some of the help documents allow you to leave feedback and give ratings. You can use this as a place to report bugs in the documentation or ask for more clarification if you still do not understand the concepts on the page.

Using Flash CS5 Help

1. Click Help.
2. Click Flash Help.

Note: *You can use the F1 keyboard shortcut to launch the Help application.*

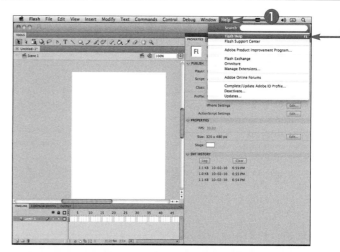

3. Click ActionScript 3.0 Language and Components Reference.

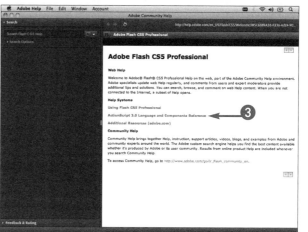

④ Select a class in the Class list, such as Accelerometer.

● You can click a property in the summary to go to its help.

● You can click a method in the summary to go to its help.

● You can click the page URL to open it in a Web browser.

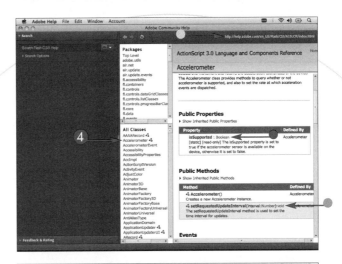

⑤ Enter a search term, such as TouchEvents.

⑥ Click a search result.

The page will be loaded in the application.

Extra

It is really important that you set any of the important Help packages to be available for offline viewing. The last thing that you will want is to find yourself without Internet access and struggling to remember how to do something. You can manage your local content from the Preferences dialog box in the application. It will also show you if there are any updates to the documents that you have on your computer. Keeping your Help files up to date is critical, as they may contain bugs just like your code can. They can also contain new features and APIs for newer versions of the Flash Player and AIR. Before the Adobe Help application, it was difficult to keep everything up to date, and you never knew if you had the most up-to-date files. Adobe recognized these issues and is committed to making the Help application as easy as it possibly can be.

Create a New Project

Adobe has created a new project template for creating iPhone applications. When you select the iPhone template in the New Document dialog box, an .fla will be created with Stage dimensions of 320 x 480, which is the full-screen resolution of the iPhone in portrait mode. Your file will also be set up to publish with ActionScript 3.0 and with the default iPhone settings.

If you are creating an application that will only be in landscape mode, you can change the Stage dimensions to 480 x 320 in the Document Settings dialog box.

In the Document Settings dialog box, you can also set the background color and frame rate of your application. The default frame rate for a new .fla file is 24 frames per second. This is the standard frame rate for animations; however, you may want to change this if you are creating a game. 30 frames per second is a commonly used frame rate for Flash games and Web sites; however, feel free to experiment with this to find what works best for you. The faster your frame rate, the faster your animations will happen. The downfall to this is that you may be redrawing your graphics more often, which may cause the performance of your application to decrease.

There is also a Make Default button in the Document Settings dialog box. Clicking this button will set the current document's settings to the new default for any new file that you create. This will affect every type of .fla and not just your iPhone projects.

Create a New Project

1. Click File.
2. Click New.

Note: *You can also use the ⌘+N (Ctrl+N) keyboard shortcut to open the New Document dialog box.*

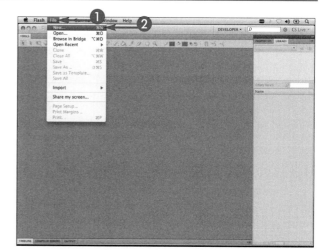

The New Document dialog box appears.

3. Click iPhone OS in the Type column.
4. Click OK.

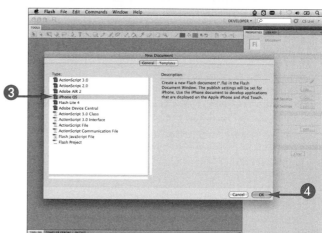

A new, blank iPhone project is created.

5 Click Modify.

6 Click Document.

Note: You can also use the ⌘+J (Ctrl+J) keyboard shortcut to open Document Settings dialog box.

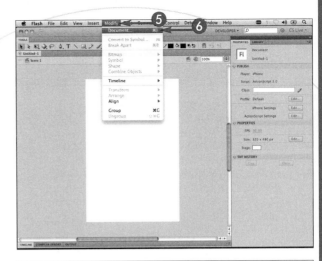

The Document Settings dialog box appears.

7 Set the Frame rate, such as 30.00 fps.

- You can change other basic document settings here as well.

8 Click OK.

Your changes are applied to the new document.

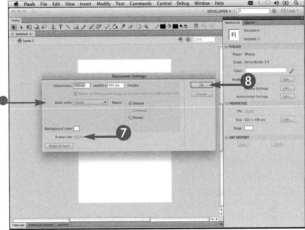

Chapter 3: Developing Your First Application

Extra

Flash CS5 has introduced a new file format for your Flash projects, .xfl. This file format is an uncompressed version of the .fla format. To save your file in this format, select Flash CS5 Uncompressed Document (*.xfl) from the Format drop-down list in the Save dialog box. This will create a folder with the same name as the filename you specified. Inside that folder is a folder structure that contains all the information about your file, as well as the assets in your Library. If you used CS5 to import any images or audio files to the Library, they can be found in the LIBRARY folder. This allows you to update assets without having to open the file or re-import them. The DOMDocument.xml file is an XML representation of your file. It contains all the ActionScript on any frames and any shapes or assets on the Stage.

The .xfl format makes adding Flash files to a source control system, such as Subversion, a lot easier. Because the file format is text, as opposed to the binary .fla file format, you will have the ability to merge changes between two versions of your files. This will make working with large teams a lot more efficient.

53

Configure Publish Settings

Selecting the iPhone template from the New Document dialog box sets up some of the major publish settings for you. However, there are many other settings that can affect your final output. You can set these in the Publish Settings dialog box. By default, there are three tabs across the top: Formats, Flash, and HTML. The Formats tab enables you to select multiple types of files to output your file to. Because you are only concerned about creating iPhone applications, you can deselect the HTML check box, which removes the HTML tab from the top.

The text input area beside the Flash check box is the location where Flash will output the compiled .swf for your application. This allows you to set the location to another folder other than the same one as the .fla. It is a good practice to use relative paths to your file just in case you copy the project folder to a new location or you are working with a team of developers and designers.

The Flash tab has all the settings that will affect the output of your .swf file. Changing the compression settings for images and sounds will greatly affect the file size and the quality of your application. You can also publish the file from the Publish Settings dialog box by clicking the Publish button at the bottom. Publishing your file will export your file to the .swf file specified on the Formats tab and create the iPhone application in the same directory.

Configure Publish Settings

1. Click File.
2. Click Publish Settings.

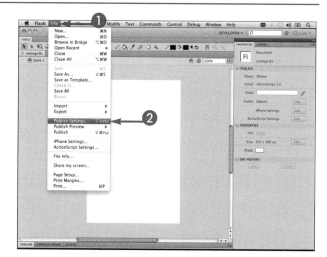

The Publish Settings dialog box appears.

3. Click here to uncheck the HTML check box.
4. Set the path to your .swf file.

Note: *The default .swf file location and name is the same folder as your .fla file with the same name.*

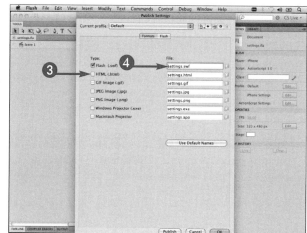

5 Click the Flash tab.

6 Set the default JPEG quality settings, such as to 100.

7 Click OK.

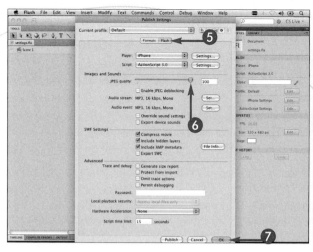

Your publish settings are applied to your document.

8 Test the movie by pressing ⌘+Enter (Ctr+Enter).

9 Check to see if your .swf file was created in the location you specified in step **4**.

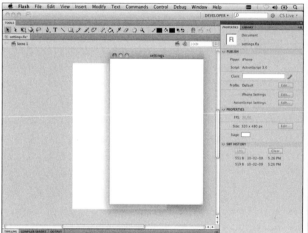

Extra

If you often find yourself adjusting the Publish settings the same way, you can save that profile to be used later. At the very top of the Publish Settings dialog box, there is a drop-down list with the currently selected profile and a set of buttons to the right of it. The first button exports and imports the publish profiles. These are exported as .xml files and can be imported from another file to change the settings. The next two buttons are Create New Profile and Duplicate Profile, which essentially do the same thing. These add a new profile to the .fla file. This comes in very handy if you are targeting multiple platforms. For example, you could create a profile that targets the iPhone, one that targets AIR, and one that targets the Web. This allows you to publish to multiple platforms quickly during development when you make changes to your application. It also gives you control over being able to optimize your assets for the different platforms. On the desktop, you will want your assets at their best quality, whereas on the Web you will probably compromise quality for file size. The fourth button enables you to rename the currently selected profile. Finally, the fifth button deletes the currently selected profile.

Set Your Application Output

There are other settings besides those in the Publish Settings dialog box that will affect your final output. You can find these settings in the iPhone Settings dialog box, which can be accessed from the Flash tab of the Publish Settings dialog box or as shown below. The General Tab contains the main settings for your application.

The output file is the location and name of the .ipa file that will be created when you publish the file. This is the file that you will install onto your device. You can name your file anything you like, but it is a good idea to keep it consistent with other output files that you create such as your .swf file.

The app name is the name that you want to give your application. This is the text that will appear underneath the icon on the home screen of your iPhone. There is limited space underneath the icon, so be sure to pick a good name. A name with 13 characters or less will fit; any more will be truncated with "..." in the middle of it.

The Version input area of the iPhone Settings dialog box is where you set the version number of your application. When you install your application to your device through iTunes, iTunes uses the version number to determine if your application needs to be updated. This means that if you make changes to your application and do not update the version number, iTunes will not recognize that the application has been updated and will not sync it to the device. You can work around this requirement by installing your application through Xcode if you are developing on Mac OS X. For more details about version numbers, see the section "Update Your Version Number" later in this chapter.

Set Your Application Output

1. Click File.
2. Click iPhone OS Settings.

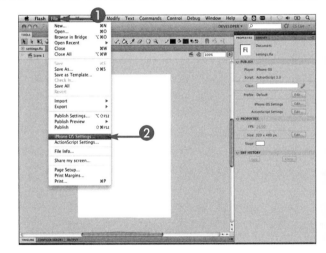

The iPhone Settings dialog box appears.

3. Change the name and location of your .ipa file.

④ Select a name for your application, such as Settings App.

Note: Try to keep this 13 characters or smaller.

⑤ Set your version number, such as 0.1.

⑥ Click OK.

Your iPhone settings are applied to your document.

Extra

When you publish your application, Flash creates a file with the .ipa extension. An .ipa file is an iOS application file and by default is associated with iTunes on your computer. However, an .ipa file is no more than a fancy zip file. To examine the file in greater detail, you can rename the file to have a .zip extension and extract it. This will extract a Payload folder with an .app file inside of it. If you have ever used Mac OS X, you will be familiar with an .app file as this is the main application file type for Mac OS X. To examine the .app file, you can right-click it, or Ctrl+click it, and select Show Package Contents in a Finder Window. This will open a new Finder window with the contents of your .app. Here you will see any of the external files that you bundled with your application. The Info.plist file is a representation of your application descriptor file and contains configuration information for your application. There will also be another .ipa file, which is the binary of your application. This is the executable that is launched when your application is started. The bigger that this file gets, the longer it will take for your application to load.

Add Your iPhone Certificate

You can add your iPhone certificate to your application on the Deployment tab of the iPhone Settings dialog box. If you have not created an iPhone certificate, follow the steps in Chapter 1, "Getting Started with iPhone Development," for the operating system that you are developing on. Without a certificate, you will not be able to compile your file to an iPhone application.

In the iPhone Settings dialog box, you can browse for the file or use the Certificate field's drop-down list. Flash saves the location of every certificate file that you enter into any Flash file. Each certificate will be populated in the drop-down list, so you can select them again easily. This enables you to easily switch between a development certificate and a distribution certificate.

A development certificate is to be used only during the development of your application. When you are ready to submit your application to the App Store, you will create a distribution certificate in the Program Portal and use that for compiling your application.

When you created your .p12 file, you were prompted to create a password. You will need to enter this in the Password input field of the iPhone Settings dialog box. If you check the Remember Password for This Session check box, Flash will remember your password until you quit Flash. When you open Flash back up again, you will need to make sure that you reenter your password before publishing. If you do not reenter your password, Flash will throw an error and halt the publishing process.

Add Your iPhone Certificate

1. Click File.
2. Click iPhone OS Settings.

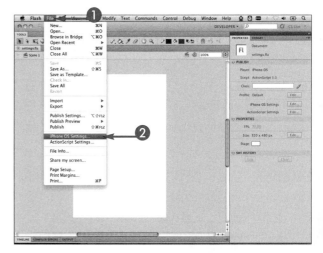

The iPhone Settings dialog box appears.

3. Click the Deployment tab.
4. Click the folder icon and browse to your .p12 file.

Note: In Windows, you click the Browse button and browse to your .p12 file.

OR

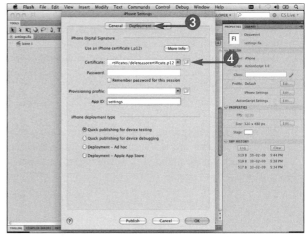

④ Click here and select a previously entered certificate.

⑤ Type the password that you entered when creating your .p12 file.

⑥ Click Remember Password for This Session.

⑦ Click OK.

Your iPhone certificate is added to your application.

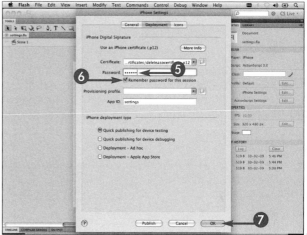

Apply It

If you are developing on the Mac OS X operating system, you can tell Flash to get the certificate information from your keychain. This will save you from exporting the .p12 file from your certificate as described in Chapter 1. This cannot be done with the Flash IDE but can be accomplished when compiling your application with the command line. For more details, see the section "Compile from the Command Line" later in this chapter. The following syntax will enable you to use the certificate in your keychain:

```
java -jar /Applications/Adobe\ Flash\ CS5/PFI/lib/pfi.jar -package -target ipa-test
  -provisioning-profile "/Users/julian/Downloads/flashdevelopment.mobileprovision" -alias
  "iPhone Developer: Julian Dolce (XXXXXXX)" -storetype KeychainStore -providerName Apple
  "settings.ipa" "settings-app.xml" "settings.swf"
```

There is currently a Java limitation with retrieving keys from the keychain if you have more than one. If this is the case, you are better off exporting the P12 certificate for now.

Add Your Provisioning File

Provisioning profiles connect your applications with a valid iOS device that is registered with your iPhone development account. If you have not created one yet, see Chapter 1 for details. Without a proper provisioning profile installed on your device, you will not be able to install the application on it.

After you have created your provisioning profile file, you can add it to your application using the iPhone Settings dialog box, which can be accessed from the Flash tab of the Publish Settings dialog box or as shown below. You can browse for your .mobileprovision file or use the Provisioning Profile field's drop-down list. Every time you add a new provisioning profile file to any file, Flash will remember and include it in the drop-down list. This enables you to quickly select frequently used profiles and saves you from having to enter in the path every time that you create a new file.

When you created your provisioning profile, you selected an App ID to associate it with. You will need to enter that App ID in the iPhone Settings dialog box. This ID distinguishes your application from another on your device, in the off chance that your application has the same name as another. If you created a wildcard App ID, replace the * with a word that will make your application unique. This will usually be something similar to the application name.

Add Your Provisioning File

① Click File.

② Click iPhone OS Settings.

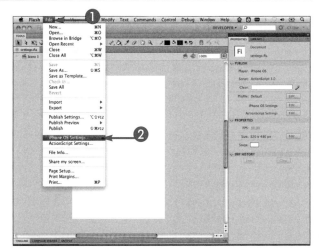

The iPhone Settings dialog box appears.

③ Click the Deployment tab.

④ Click the folder icon and browse to your .mobileprovision file.

Note: *In Windows, you click the Browse button and browse to your .mobileprovision file.*

OR

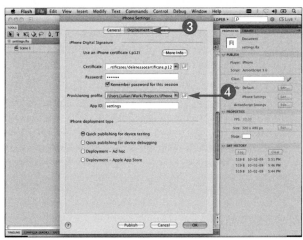

④ Click here and select a previously selected .mobileprovision file.

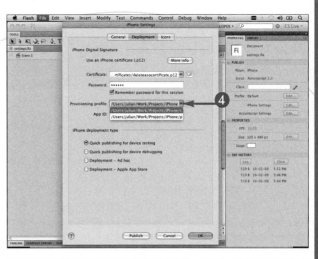

⑤ Enter an App ID that is included in your provisioning profile, such as com.deleteaso.settings.

⑥ Click OK.

Your provisioning profile is added to your application.

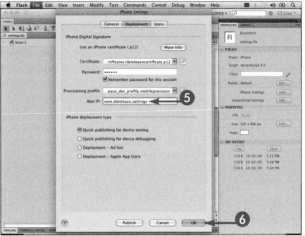

Extra

There are different types of provisioning files, and you will need to select the appropriate one for the type of application you are publishing. Chapter 1 shows how to create a development profile, which is to be used during the development of your application. Only developers who are included in your iPhone developer program will be able to install applications that use a development profile.

When you are ready to have nondevelopers test your application or to submit the application to the App Store, you will need to select a different type of provisioning profile for that specific action — a distribution profile. When you send your application to a nondeveloper, you will also need to send her the proper provisioning profile. She can install it on her device by adding it to her iTunes Library. This will cause the file to be installed the next time she syncs her device.

Compile from Flash Professional CS5

After you have created an iPhone file in Flash Professional CS5 and properly set all the publish settings, you are ready to compile the file to an iPhone application. If you have not created the necessary iPhone developer files, see Chapter 1 for more details and the early topics in this chapter in order to properly add them to your file. This is only necessary if you have to publish an .ipa file, which you can install on a device.

You can test your application on your computer; however, some of the features will not work, as they are available only on your device — for example, the accelerometer and geolocation feature. The entire process to compile and install an application on your device can take up valuable development time. If you can test your application, even in parts, on your computer before installing it to your device, this will save you lots of time. After you are satisfied that your application works on your computer, install and test it on a device because it will run differently than on your computer.

To test your application, follow the steps below to compile and launch it in the AIR debugger, in which you will be able to receive any trace statements in the Output window.

When you are ready to create your application to install it on your device, publish it as shown below. This will take a few moments, so be patient as Flash creates your .ipa file.

Compile from Flash Professional CS5

Test Your Application

1. Click Control.
2. Click Test Movie.
3. Click In AIR Debug Launcher (Mobile).

Note: You can also use the ⌘ *(Ctrl)+Enter keyboard shortcut to test your movie.*

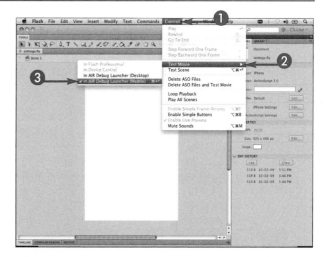

Your application is compiled and launched on your machine.

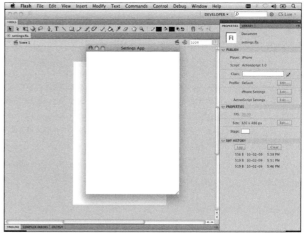

Publish Your Application

1. Click File.
2. Click Publish.

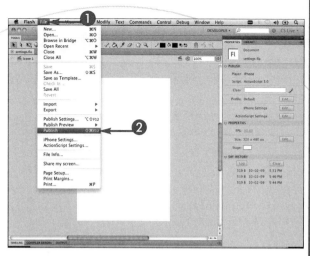

- An .ipa file is created.

Chapter 3: Developing Your First Application

Extra

When you are testing your application, be sure to check the Compiler Errors panel. If there is a problem with any of the ActionScript you have written, the Flash compiler will stop and throw an error, which is displayed in the panel. Each error will give you the location and line number of the error, as well as a description as to what the error is. Clicking the error in the panel will bring you to the line number and position of the error in your code.

The Compiler Errors panel shows warnings as well as errors. These warnings are not severe enough to cause Flash to not compile your file; however, it is probably a good idea to investigate each one and try and resolve them because they may cause unexpected behavior in your application. Some examples of the errors you may encounter are duplicate variable names, duplicate frame labels, and AS3 migration issues.

In Flash CS5, you can now toggle the visibility of the errors and warnings by clicking the representative icons at the bottom of the Compiler Errors panel.

Compile from the Command Line

The iPhone Packager is a Java application that creates your iPhone application from an .swf file. This is the application that Flash CS5 uses when compiling from it. You can also use this application to compile your applications from a command line. The pfi.jar Java application can be found in the PFI/lib folder in the Adobe Flash CS5 installation folder.

Before you start, you will need to get the full paths to the pfi.jar file, your certificate .p12 file, and your mobile provisioning file. After you have the paths to those files, open a command shell or a Terminal window and navigate to the location of your .swf file.

The iPhone packager has a number of switches that are used to configure how your application is compiled. The -target switch allows you to select which type of application you want to compile. You have the same options that are available on the Deployment tab of the iPhone Settings dialog box in the Flash IDE. The -provisioning-profile switch enables you to set the mobile provisioning file to use. The -storetype switch specifies the type of certificate you are using, which in this case is pkcs12. The -keystore switch specifies the file to your .p12 certificate file. -storepass is the password for your certificate file.

The parameters are as follows: the path to the .ipa file you want to compile, your application descriptor file, and the .swf file of your application. These are all the required parameters and switches to compile your iPhone application. If you have included any paths to icons or default images in your application descriptor file, you can add these after the required fields.

Compile from the Command Line

① Open a Terminal or Command window.

② Navigate to the folder with your .swf file, such as cd /Users/julian/Desktop/settingsproject, and press Enter.

③ Type **java –jar**.

④ Type the path to the pfi.jar, such as "/Applications/Adobe Flash CS5/PFI/lib/pfi.jar", and press Enter.

⑤ Type **-package –target**.

⑥ Enter a valid -target option, such as ipa-test.

⑦ Type **-provisioning-profile**.

⑧ Type the complete path to your provisioning file, such as "deleteaso_dev_profile.mobileprovision".

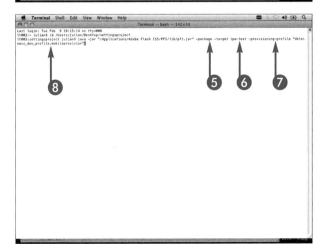

⑨ Type **-storetype pkcs12 –keystore**.

⑩ Type the full path to your .p12 file, such as "deleteasocertificate.p12".

⑪ Type **–storepass**.

⑫ Enter your certificate password, such as mypassword.

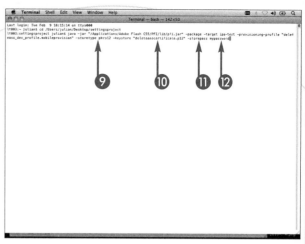

⑬ Type the path where you want your .ipa file compiled, such as "settings.ipa".

⑭ Type the full path to your application descriptor file, such as "settings-app.xml".

⑮ Type the full path to the .swf file you want to compile to an .ipa file, such as "settings.swf".

⑯ Press Enter.

Your application is compiled.

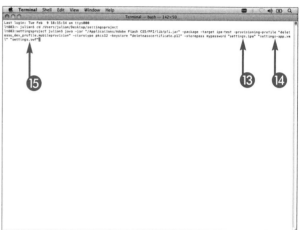

Apply It

There are four valid values for the `-target` switch, which specifies the iPhone deployment type.

Use the following `ipa-test` option if you want to quickly test the application and install it on your device:

```
-target ipa-test
```

If you want to debug your application with the Flash Debugger, you can use the `ipa-debug` option. This will allow you to receive `trace` statements in the Output panel. For more details on this, see Chapter 17, "Debugging Your Application."

```
-target ipa-debug
```

To compile your application for ad hoc deployment, use the `ipa-ad-hoc` option:

```
-target ipa-ad-hoc
```

To compile your application for submission to the iTunes App Store, use the `ipa-app-store` option:

```
-target ipa-app-store
```

Remember to use the proper provisioning files and certificates when you are doing development versus distribution compilation.

Install Your Application with iTunes

When you compile your application to an iPhone application, an .ipa file is made. By default, .ipa files are associated with iTunes, and double-clicking them will add them to your Applications section. Before installing your application, make sure that you have the most up-to-date version of iTunes.

The Applications section in iTunes has all the applications that you have downloaded or purchased from the App Store. You can also drag your .ipa file onto the list of applications to add it to your Library.

When you add your application to the iTunes Applications Library, it will be installed to your device the next time that you sync it with your computer. Syncing your device with iTunes initiates a few events besides simply installing your application. First, it will create a backup of your device, as it does most times when you sync your device. Depending on how much data you have on your device, the time for this to be completed will vary. After the backup is complete, iTunes will perform a full sync of your device. This will not only install your application to your device, but also add any music and videos that need to be added as well. This can potentially cause very long sync times with your device.

These long wait times can be annoying if you are trying to test your application on your device and quickly make changes and updates. Installing your application with iTunes is not the best method for installing application during development; however, it will be the easiest for nondevelopers when they are testing or trying your application. It also does not require any additional downloads because if someone has a device you can almost guarantee that they have iTunes already installed on his or her computer.

Install Your Application with iTunes

1. With your device attached, in iTunes, click File.
2. Click Add to Library.

   The Add To Library dialog box appears.
3. Navigate to and select the .ipa file that you want to install on your device.
4. Click Choose.

   You are returned to the iTunes window.
5. Click Applications under the Library heading.
   - Your application is added to the Library.

6. Click your device's name.
7. Click the Applications tab.
8. Make sure that your application is check in the Sync Applications list.
9. Click and drag your application's icon to change its position on your phone.

10. Click Sync.
- The Status area will show the sync progress.

Extra

If you decide to use iTunes as your method to install your applications on your device, there are a couple of items to keep in mind. When updating your application, you may not see the changes that you made. This is caused by not updating the version number of your application, as mentioned earlier. If you do not update it, iTunes will think that the two files are the same version of the application and not sync it to your device, even though the file has actually changed. It is a good practice to always update the version number as you develop, but in case you want to quickly prototype or try out some ideas, it can be an annoying extra step. It is also a good idea to visually show the version number in the interface. This way you will know for sure exactly what version you have installed on your device. For more details on updating your version number, see the section "Update Your Version Number" later in this chapter. For better development options, use the iPhone Configuration Utility and Xcode to install your application, which are explored in the following two sections.

Install Your Application with the iPhone Configuration Utility

The iPhone Configuration Utility is an application that runs on both Windows and Mac OS X and allows you to easily maintain and set up devices. Originally created to help manage devices in an enterprise environment, the iPhone Configuration Utility has some great features that will help developers be more efficient.

The iPhone Configuration Utility can be used to install applications on your device. However, it does not support .ipa files, and you must convert them to a file with the .app file extension. To accomplish this, simply rename the .ipa file to have a .zip file extension and extract it. This should create a Payload folder with your .app file inside of it.

When you use the iPhone Configuration Utility, it adds your application to the Library for the utility just like when you add it to your iTunes Library.

Before you can install the application to your device, remember to connect it to your computer. When your device is selected in the iPhone Configuration Utility, under the Applications tab, you will see a list of all the apps you have currently installed on your device, along with any apps that you have added to the Library.

The Applications list enables you to perform actions on specific applications. You will see an Uninstall button beside any of the applications that you have developed and an Install button beside any applications that have been added to the Library. These buttons save you from having to sync your entire device every time that you want to add your application to your device.

Install Your Application with the iPhone Configuration Utility

Create an .app File

1. Change the name of your .ipa file to a .zip file, such as settings.zip.
2. Double-click the .zip file to extract it.
3. Expand the Payload folder and find your .app file.

Install Your Application

1. With your device attached, open the iPhone Configuration Utility application.

Note: The utility can be found in the iPhone Configuration Utility folder in your Program Files folder in Windows and in /Applications/Utilities on Mac OS X.

2. Click Applications.
3. Click Add.

   The Add dialog box appears.

4. Click your .app file.
5. Click Open.

- Your application is now added to the Library.

6. Click your device's name.
7. Click the Applications tab.
8. Find your application and click Install next to it.

   You application is installed on your device.

Extra

The iPhone Configuration Utility is also a great way to install an application on multiple devices. If you have multiple devices connected to your computer at any given time, select them to see the list of applications installed on them. Each application that is included in the iPhone Configuration Utility Library will be shown in each list, and you will have the ability to install and uninstall each. Testing on multiple devices is extremely important, as each generation of a device is faster. If you always test your application on an iPhone 3GS, you will probably see great performance. However, testing it on an iPhone 3G will show a decrease in performance.

Nondevelopers who want to install applications on their devices can also use the iPhone Configuration Utility. This is a convenient way for your testers to get your application on their devices. It is also a great way for them to see exactly what version they have installed on their devices, which is something iTunes does not show. This will help them log bugs to a specific version number and will also be useful in determining if they have the latest build.

Install Your Application with Xcode

Xcode is the primary IDE that comes bundled in the Apple developer tools and the iPhone SDK, and it is available only on Mac OS X. Xcode provides the ability to develop in a number of programming languages, as well as author iOS applications using the iPhone SDK.

The Organizer window in Xcode provides some convenient workflows for managing your applications and devices. It lists all the devices that you have added to use for development, similar to iTunes and the iPhone Configuration Utility. Selecting your device will present you with the Summary tab.

The Summary tab consists of three main locations: device information, Provisioning, and Applications. The Applications area contains a list of all the applications installed on the device. You can install an .app version of your application and make it appear in this list. To create an .app version from your .ipa, simply rename the .ipa file to have a .zip file extension and extract it. This should create a Payload folder with your .app file inside of it.

You can also drag and drop your .app file from a Finder window onto the Applications list to install it. Adding your application to the list automatically installs it on your device. You will see the progress of all the files being sent to your device right above the Provisioning section. If you have any issues with installing the application, you can monitor this area to see where in the installation process it fails. This will help narrow down any issues.

The best thing about installing applications through Xcode is that they do not require their version number to be updated every time. This will save you lots of time during development.

Install Your Application with Xcode

Create an .app File

1. Change the name of your .ipa file to a .zip file, such as settings.zip.
2. Double-click the .zip file to extract it.
3. Expand the Payload folder and find your .app file.

Install Your Application

1. With your device attached, open Xcode.

   Note: Xcode can be found in the folder /Developer/Applications.

2. Click Window.
3. Click Organizer.

The Organizer window appears.

4 Click your device's name.

5 Click the + button.

A dialog box appears.

6 Click your .app file.

7 Click Open.

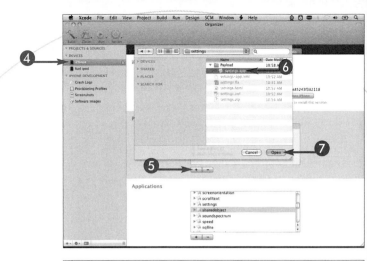

- Your application will automatically begin to install on your device.

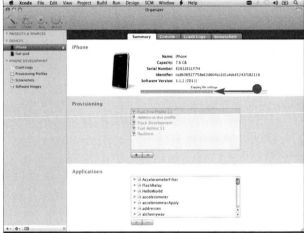

Extra

If you are developing on an Apple computer running Leopard or Snow Leopard, you will probably use Xcode for all your application and device management. It provides all the same features that iTunes and the other applications explored earlier in this chapter do, plus more. The only downfall is that you have to convert your .app file to a .zip file every time. Although this process is straightforward and does not take a lot of time, it is just one more thing that you have to do to get your application on your device.

There are many different ways that you can automate this. If you are compiling your application from the command line, you are probably already using some sort of batch script. You could easily have the script rename the .ipa file that it outputs and extract it after it has completed publishing. You can also use the Automator application that comes with Mac OS X to create a workflow that will do this. You can set the Auotmator workflow to monitor a folder and watch for any .ipa files, and when it finds one, it will rename it to a .zip.

Update Your Version Number

Setting the version number helps you distinguish between different versions of your application. When installing your application with iTunes, you must update this number every time. A common number scheme is major.minor.build. The major number is incremented when you release a version to the App Store, the minor number is incremented when you release a new build to testers or a small set of features for its major release has been completed, and the build number gets incremented every time you publish the file during development. When a number is incremented, also set the number to the right of it to 0.

You can set the version number in two places. The first, and easiest, is on the Deployment tab in the iPhone Settings dialog box. The version number input area is directly under where you added the name for your application.

The second place is in the app descriptor file that is created when you publish your application. This .xml file is created in the same directory as your .swf file and is named <swfname>-app.xml. To edit the version, open the file in Flash CS5 or any other text editor and find the <version> node. Simply replace the current version with the new one and save the file.

Being able to update the version from this file allows you to write a batch script to automatically update the version for you. There are several ways to create this script, depending on the workflow you have set up in order to compile your applications. One way would be to write a JSFL script that could do this. If you are unfamiliar with JSFL, check the Flash Help for more details.

Update Your Version Number

Update the Version Number in Flash

① Click File.

② Click iPhone OS Settings.

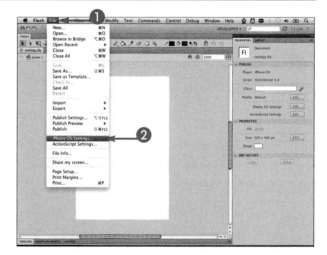

The iPhone Settings dialog box appears.

③ Increment the version number, such as 1.1.0.

④ Click OK.

The version number is updated.

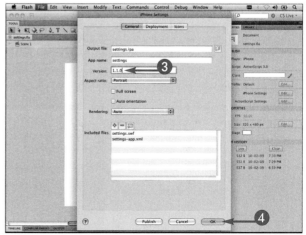

Update the Version Number in the App Descriptor File

1 Click File → Open.

The Open dialog box appears.

2 Click your application descriptor file, such as settings-app.xml.

3 Click Open.

Your application descriptor file opens.

4 Update the version number in the `<version>` node, such as 1.1.0.

5 Click File → Save.

The file is saved with the updated version number.

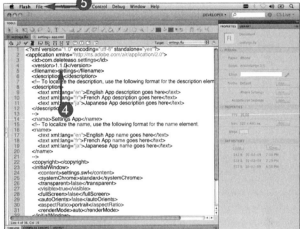

Apply It

When you are sending builds to your testers, it is a good idea to display the version number somewhere in the interface. This will allow them to log bugs to a specific version and will prevent any confusion on which build they actually have on their phones. Here is the syntax to retrieve the version number with ActionScript:

```
var appDescriptor:XML = NativeApplication.nativeApplication.applicationDescriptor;
var ns:Namespace = appDescriptor.namespace();
var appVersion:String = appDescriptor.ns::version;
```

Explore Apple's Human Interface Guidelines

If you have ever spent time using the Mac OS X operating system, you have probably realized that many of the applications have a similar look and feel. This allows users to become familiar and comfortable with applications quickly. In order to help developers with this, Apple has created a set of useful controls and components that are beautifully designed. To match its efforts with desktop applications, Apple has done the same for iPhone apps. If you have used enough apps, you can start to pick out which elements are part of the iPhone SDK. Currently, there is not a way to incorporate the native controls with your Flash CS5 application; however, there are many Photoshop templates online that will allow you to make your own controls look like Apple's. If you decide to use buttons and graphics that look like the ones provided in the SDK, make sure that they behave the same way that Apple had intended. If you implement a button or graphic that does not do what Apple designed it to do, there is a good chance that Apple will reject your application when you submit it to the App Store. You will explore the intention of various graphics and icons in Chapter 6, "Working with Images."

One Screen

One of the biggest differences between designing for the desktop and mobile platforms is the amount of windows that your application can have. On the desktop, an application can theoretically have as many windows that it needs. On a mobile device, however, you are able to show only one screen at a time. There are exceptions to this rule, such as the different alert and modal windows, but plan for your application to have only the one window. Your application can have multiple screens, but only one can be shown at a time. If your application requires the use of multiple windows, you will need to try and design it into a sequence of screens or rethink the user flow of your application.

One Application at a Time

With any mobile device, memory is going to be limited, and the iPhone and iPod touch are no different. Developers will need to make sure that their applications use the least amount of memory as possible. To stop applications from using memory when they are no longer being used, Apple has limited the number of applications that you can run at the same time to one. This means that a third-party application will never run in the background after a user has exited it. So when a user answers the phone or switches to any other application, the application that he or she was using will quit.

If a user is interrupted unexpectedly while using your application, he will expect that his data is not lost when he returns. To help combat this, it is important to save the state of your application when it makes sense. You may also want to save the last screen or state in which the user left the application. When the user returns to the application, you can load the necessary data and take him to the last part of the application that he was interacting with.

For more details on the several different ways to save data and states in your application, see Chapter 10, "Saving State."

Think Top Down

When designing your application, you will need to take into consideration how you display information to the user. Because a user will be using her fingers or thumbs in order to interact with your application, portions of the screen will be blocked from view by her hand. This is an important user experience problem that designers and developers will need to solve as touch-enabled devices and screens become more ubiquitous. It is a good practice to place any important information higher up on the screen than you normally would. Any changes in your interface as the user interacts with it should always be above where she is touching the screen. This will make sure that the user's hand does not block any important information.

Minimize Input

One of the big differences between the iPhone and other smartphones is the keyboard. Other smartphones have a tactile keyboard that is integrated into the hardware of the device. With the screen size of the iPhone, adding a real keyboard would not be practical. Instead, there is a touch-screen keyboard that appears when user input is required. Typing on this keyboard is suitable for typing short amounts of text in small bursts, but this is less than ideal when trying to type a long email. Having users enter lots of information with the keyboard can cause them to become frustrated and leave your application. Limiting the amount of information that users must input before having something meaningful occur is a good way to keep them happy. It is also a good idea to make users enter that information only once. Saving it to the device to a text file or `SharedObject` is an option for doing this. You can also create a settings bundle for your application so that users can update this information from the Settings application on the device. This way, if any of their information ever changes, they can update it without going into the application. For more details on how to create settings for your application, see Chapter 16, "Creating Application Settings."

Focus

Users' typical interaction with their iPhones comes in short bursts when they are not at their computers. Because of this, their attention span will be extremely short; they will want the information that they are looking for quickly. Keeping the focus of your application to a simple task will give you a better opportunity to effectively communicate with your users. Too much information or too many features can complicate the user experience of an application quickly. For example, using an application to check up-to-date sports scores probably loses focus if the user can purchase tickets for the game as well. These two features would probably serve your users better if they were separate applications.

Keeping your application focused on a single task will reduce the amount of help you need to provide to your users. When you present the users with a screen, they should not have to ask themselves what they are supposed to do. Because screen real estate is at a premium, you should avoid having large pieces of text explaining what to do. If you find yourself having to explain your app's functionality to the user, chances are it is too complicated. It is good to discover this as early in development as possible. Creating storyboards and mock-ups will allow you to plan effectively. Have an unbiased party look these over and see if he or she understands the goal of your application. The more planning and focus group testing that you can do upfront, the more time you will save in development.

Understanding Screen Resolutions

When designing your application to run on the iPhone, it is extremely important to factor in its compact screen size. Creating a device that is portable and fits in your pocket can create many design challenges. Designing for the 3.5-inch diagonal screen size of the iPhone is much different than designing for a 19-inch desktop monitor.

The iPhone has a screen resolution of 480 x 320 while in Portrait mode and 320 x 480 while in Landscape mode. If you design your application to have the status bar visible, you will need to account for it during design. The status bar is 20 pixels high and sits on top of your content. This means that some of your content could be hidden from the user if this is left unaccounted for. If the status bar is not a required element of your application, you can create a full-screen application that will hide it from the user. See the following section, "Create Full-Screen Applications," for more details.

Considering screen real estate when designing your application will force you to make sure that you include only necessary elements for interacting with your application. Crowding the interface with unnecessary design elements can confuse your users and provide a bad user experience. Every element on the screen should have a purpose, whether it is displaying information to the user or allowing him or her to interact with it.

Branding should be used cautiously and incorporated in subtle and unobtrusive ways. Keep branding focus to the application icon on the home screen and the default loading screen.

Examples of Applications in Portrait Mode

The following application is 460 x 320.

A The content height is 460 pixels.

B The content width is 320 pixels.

Examples of Applications in Portrait Mode *(continued)*

The following application is 480 x 320.

- (A) The content height is 480 pixels.
- (B) The content width is 320 pixels.

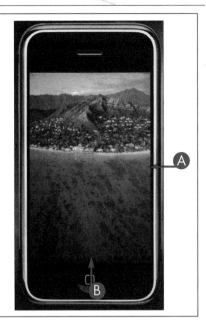

Examples of Applications in Landscape Mode

The following application is 320 x 480.

- (A) The content height is 320 pixels.
- (B) The content width is 480 pixels.

The following application is 300 x 480. Notice how the status bar is over the top of the content.

- (A) The content height is 300 pixels.
- (B) The content width is 480 pixels.

The iPad's Screen Resolution

Apple has introduced us with a device that has a bigger screen resolution, the iPad. The iPad has a 9.7-inch diagonal screen and a screen resolution of 1024 x 768. If you design your application at the iPhone screen resolution, it can still run at full screen on the iPad. However, that would not be taking advantage of the technology. If you plan on supporting both devices, or multiple platforms, consider how your application would readjust its layout based on a new screen resolution at runtime. You can easily test how your application would respond to screen size changes by resizing the window when testing on your computer.

Of course, this goes beyond the iPad. Thinking about how your application would look on any sized platform early on is never a bad idea. Even if you plan on never supporting other platforms today, you never know what the future holds. iPhone developers never thought they would be converting their apps to the iPad a year ago.

Create Full-Screen Applications

The status bar is the top bar on your device. It contains some important information, such as battery life, the time, cell signal, cell carrier, and the currently connected network interface. This can be important information for some applications but unnecessary for others. The status bar also is 20 pixels high and sits on top of your content, which reduces the screen resolution.

If you are creating an application that needs the extra screen real estate and does not need the status bar, you can hide it. Most games hide the status bar because they want to fully immerse the user in the experience of playing the game.

To hide the status bar to create full-screen applications, you use the iPhone Settings dialog box. When you open it, the General tab should be selected by default; however, if it is not, select it to find the Full Screen check box. Leaving this option unchecked will create your application with the status bar, and checking it will hide it.

Included with every application is an Info.plist file. This file is a property list file that contains configuration settings about your application, and it is created during compilation and bundled with your application. When you select the Full Screen check box, Flash adds the UIStatusBarHidden key with a value of true to this file. When iOS loads your application, it looks at the different keys in the Info.plist file and executes commands based on their values — in this case, hiding the status bar.

Create Full-Screen Applications

Set Your Application to Full Screen

1. Click File.
2. Click iPhone OS Settings.

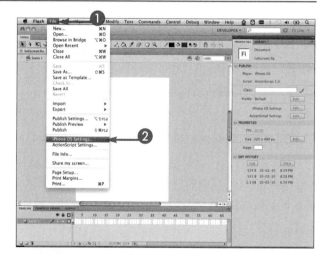

The iPhone Settings dialog box appears.

3. Click here to check the Full Screen check box.
4. Click OK.

Your application is set to run full screen.

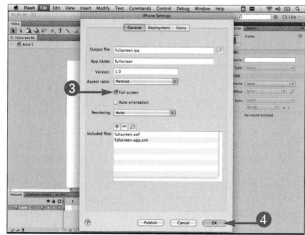

Check Your App Descriptor File's <fullScreen> Node

1. Click File → Open.

   The Open dialog box appears.

2. Click your app descriptor file.

3. Click Open.

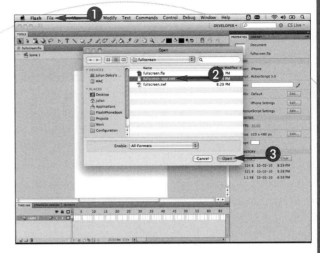

4. Find the `<fullScreen>` node.

Note: Notice that it is now set to true.

When you compile your application, the status bar will now be hidden.

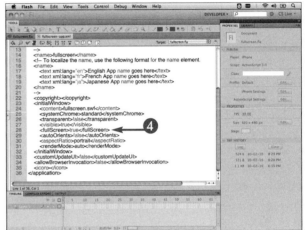

Extra

Currently, you are not able to show and hide the status bar with a Flash CS5 application. You may have noticed some applications on your device that do this while your application is running. For example, the Photos application will hide all of the UI, including the status bar, when you tap a photo. This allows the user to see the photo full screen and only shows the UI when necessary. When the user taps the screen again, the UI and status bar reappear, allowing the user to continue interacting with the application.

With this in mind, make sure that the user will never need the status bar when creating full-screen applications. If you choose to hide it, the user will have to quit out of your application to check for any of the information that the status bar holds. It is probably a good idea to display the status bar if your application does a lot of communicating with a network because your user will want to see if his or her data is being sent over WiFi or 3G.

Understanding Screen Orientation

The iPhone and iPod come equipped with three-axis accelerometers. iOS uses the accelerometer to determine in which orientation you are holding the device. As the user rotates the device in his or her hands, you can adjust your content for him or her.

You can have your application start in one of two different orientation modes, Portrait or Landscape. On the iPhone Settings dialog box's General tab is an Aspect Ratio drop-down list. This is where you can specify which orientation your application starts up in.

You can also have your content auto rotate to the correct orientation by selecting the Auto Orientation check box on the General Tab. When the device rotates your content, you may need to adjust it. To listen for orientation changes, you can add an event listener on the Stage for a `StageOrientationEvent.ORIENTATION_CHANGE` event. This event will tell you the new orientation as well as the previous one.

If you choose not to have your content rotate automatically, you can manually set the orientation. The `Stage.setOrientation()` method allows you to specify one of four valid orientations. These valid orientations can be found in the `StageOrientation` class: `StageOrientation.DEFAULT`, `StageOrientation.ROTATED_LEFT`, `StageOrientation.ROTATED_RIGHT`, `StageOrientation.UPSIDE_DOWN`.

If you plan on supporting multiple devices, you can also check to see if your device currently supports Stage orientations. The `Stage.supportsOrientationChange` is a static property that returns true if your device is able to change its orientation.

Understanding Screen Orientation

Note: *Remember, you can find the code samples throughout the book, such as screenorientation.fla here, on the Wiley Web page for the book, www.wiley.com/go/iosappsvisualblueprint, on the Downloads tab.*

Set Auto Orientation

1. Click File.
2. Click iPhone OS Settings.

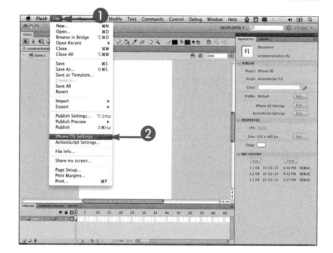

The iPhone Settings dialog box appears.

3. Click here to check the Auto Orientation check box.
4. Click OK.

   Your application is set to rotate automatically with the device.

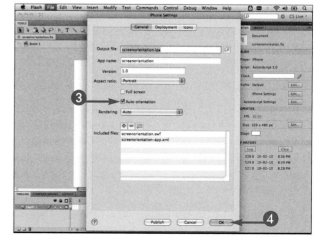

Create a Text Field to Display the Orientation

1. Click the Text tool.
2. Create and select a text field on the Stage.
3. Click here and select Classic Text.
4. Click here and select Dynamic Text.
5. Give the text field an instance name, such as `orientation_txt`.
6. Click here and select a font family, such as Helvetica.

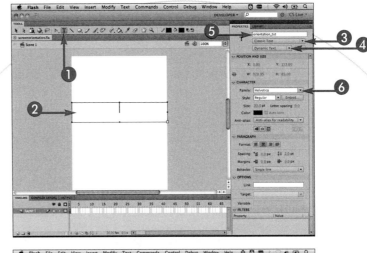

Display the Orientation in the Text Field

7. Open the Actions panel.
8. Listen for the Stage to change orientations.
9. Create an event handler function.
10. Set the text of your `TextField` to the orientation of the Stage.
11. Vertically center the `TextField`, such as `orientation_txt.y = Math.round((stage.stageHeight/2) - (orientation_txt.height/2));`.
12. Compile and install the application on your device.
13. Rotate your device to see the orientation change.

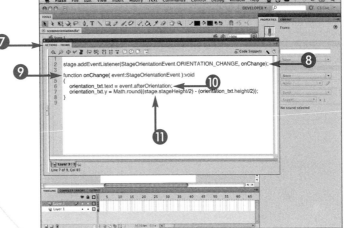

Extra

If you plan on supporting both aspect ratios, Apple recommends that you always start your application in Portrait mode. If your application supports Landscape mode, make sure that it supports both modes. If you set your application to start in Landscape mode, it will start with an orientation of `StageOrientation.ROTATED_RIGHT`. This means that the bottom of the device, the one with the connector input, will be in your right hand. Supporting the `StageOrientation.ROTATED_LEFT` orientation may make a difference for left-handed users. Also, when a user is holding the device in Landscape mode, his or her hands will be covering the speakers. This may make the audio very hard to hear, so in this case, you should not rely on sounds as a way of communicating to your user.

To create your application in Landscape mode, you can change the Stage dimensions of your Flash project to be 320 x 480. This will allow you to properly lay out your content in the correct orientation when designing your application. With the Stage selected, you can change this in the Document Settings dialog box by clicking the Edit button next to the screen dimensions in the Properties panel.

Chapter 4: Designing Your Application

Create Usable Hit States

Using a mouse as a means of interacting with an application gives you pinpoint accuracy when trying to click or interact with elements on the screen. An item that is 1 x 1 pixel in size can be clicked on, although this may be difficult. When designing touch-enabled applications, on the other hand — when the user is using his or her finger as an input device — you must take into account the size of your interactive controls.

Designing items too small or placing them too close together will cause a user to spend extra time and attention trying to tap the correct element. Additionally, there will be a lot of times when users will be using only one hand when interacting with your application. When a user is only using one hand, there is a good chance that her thumb is doing all the tapping, which will cause her taps to be less accurate.

It is important to give your users a big enough target in order for them to be less accurate with their touches. A good rule is to make sure that the target area for a touch is no smaller than 44 x 44 pixels. Depending on your application, every situation is going to be different, and it will be up to you to make sure that your elements are at an appropriate size.

It is also a good idea to provide visual feedback to the users when they tap the screen. Creating a highlighted state for your elements is a great way to do this. This will let the users know that their touch registered. Without any visual feedback, the users may think that their touch did not register if the interface does not immediately reflect their touch. For more details on how to create proper visual states for your buttons, see Chapter 5, "Handling Interaction."

Create Usable Hit States

Create a Symbol

1. Click a shape tool, such as the Rectangle tool.
2. Draw the shape on the Stage.
3. Click the Selection tool.
4. Select the shape.

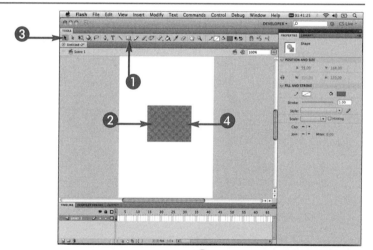

5. Click Modify.
6. Click Convert to Symbol.

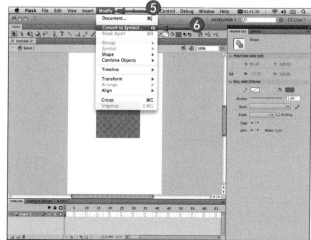

The Convert to Symbol dialog box appears.

7 Give your shape a name, such as `mybutton_btn`.

8 Click here and select Button as the type.

9 Click OK.

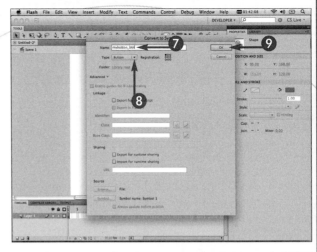

Set the Symbol's Hit States

10 Double-click the shape on the Stage to enter Edit mode.

11 Select the Up frame and press F6 three times to insert new keyframes.

12 Make sure that the Hit frame has a shape big enough to detect touches.

Note: The contents of the Hit frame do not appear when publishing your file. They are only used to determine the hit area of your button.

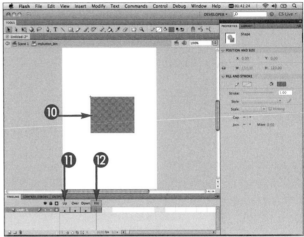

Extra

In order to try and compensate for the accuracy of your finger, iOS takes into account a certain offset from where your finger actually is. This also happens because when your finger taps an item, it will most likely cover it up. So to account for this, the iPhone will offset the hit target a little lower than you may be expecting. If your application supports only being right side up, in any orientation, then you will not have to worry about this.

However, there are certain situations in which this can come into play. Imagine if you where creating a board game such as chess or another two-player game. You could have your device on a table with each player on either side of the device where their game pieces are located. For the user who is trying to tap elements on the screen when the device is upside down for them, they will notice that they have to touch a lot higher on the item than usual. The way to get around this is to change the orientation of device when it is the other user's turn to interact.

Understanding Layout

Apple has provided some general guides to follow when laying out content in your application. When creating applications with the iPhone SDK, you lay out all your content in Interface Builder, which is part of the iPhone SDK toolset. When dragging elements on the Stage in Interface Builder, guides will appear that give you recommendations on where to place your objects.

There is not an equivalent to this in the Flash authoring environment; however, the following are some suggestions to take into consideration. Elements on the screen should not be placed against the edge of the screen and should have a 20-pixel padding from any edge. Any background images are exempt from this rule. Elements should be no less than 8 pixels apart from each other. Placing items too close to each other can cause users to tap on incorrect elements. These are just some guidelines to take into consideration, and these can be broken if there is a need.

Supporting multiple screen sizes will most likely mean that you will need to reposition items when the screen resolution changes. You can listen for the `Event.RESIZE` event on the Stage to determine whether the screen resolution has changed. In your event handler, you can determine what the new dimensions are and adjust the position of your items to their new locations. Making sure that the Stage is aligned to the top left will make calculating your content's new positions a lot easier. To set the Stage alignment to the top left, you can use the following code:

```
stage.align = StageAlign.TOP_LEFT;
```

In the example shown here, you will create an application that auto rotates and keeps a symbol centered on the screen.

Understanding Layout

Set Auto Orientation

1. Click File → iPhone OS Settings.

    The iPhone Settings dialog box appears.

2. Click here to select the Auto Orientation check box.

3. Click OK.

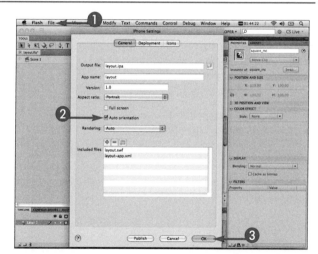

Create a Symbol

4. Create a `MovieClip` and select it on the Stage.

5. Give it an instance name, such as `square_mc`.

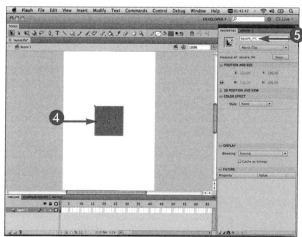

Set the Stage Properties

6 Open the Actions panel.

7 Set the Stage alignment, such as `stage.align = StageAlign.TOP_LEFT;`.

8 Set the Stage scale mode, such as `stage.scaleMode = StageScaleMode.NO_SCALE;`.

9 Add a listener for Stage resize events, such as `stage.addEventListener( Event.RESIZE, onStageResize );`.

10 Create an event handler function for your listener, such as `onStageResize`.

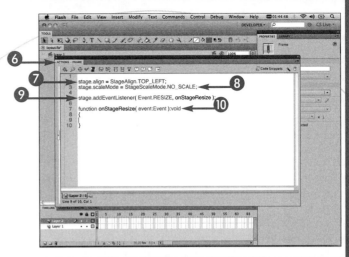

Keep the Symbol Centered

11 Center your `MovieClip` on the x-axis based on the new Stage width, such as `square_mc.x = (stage.stageWidth/2) - (square_mc.width/2);`.

12 Center your `MovieClip` on the y-axis based on the new Stage height, such as `square_mc.y = (stage.stageHeight/2) - (square_mc.height/2);`.

Note: *These equations assume the registration point of your symbol is at 0,0.*

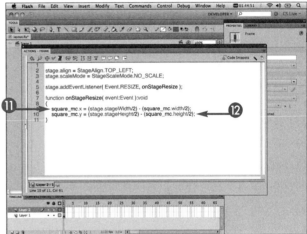

Extra

There are a number of different ways that you can align the Stage, instead of to top left. The `StageAlign` class provides seven other options for Stage alignment. `StageAlign.TOP` centers the content horizontally and vertically and aligns it to the top of the screen. `StageAlign.BOTTOM` centers the content horizontally and vertically and aligns it to the bottom of the screen. `StageAlign.LEFT` horizontally aligns the content to the left and vertically centers it. `StageAlign.RIGHT` horizontally aligns the content to the right and vertically centers it. `StageAlign.TOP_RIGHT` aligns your content to the top and right sides of the screen. `StageAlign.BOTTOM_LEFT` aligns your content to the bottom and left sides of the screen. `StageAlign.BOTTOM_RIGHT` aligns your content to the bottom and right sides of the screen. I encourage you to experiment with all of the different Stage alignments to see how your content is repositioned when the screen is resized. To initiate a resize event, you can simply drag the bottom corner of the window when testing locally on your computer.

Change the Status Bar Style

The default color for the status bar is gray. Apple has provided two additional skins that are black for the status bar in case the gray does not look good with your color palette. The black skins are almost the same except one is translucent with an alpha of 50%.

The status bar color can only be changed when your application first launches. To set the color, you add a new key to the Info.plist file, which gets bundled with every application and has a number of settings that are set when your application launches.

To add keys to the Info.plist file, you add them to your application descriptor file. This is the .xml file that Flash creates that describes your application, and it is named with the syntax *swfname*-app.xml. This file can be found in the same directory as your .swf file.

The application descriptor file contains an xml node called <iPhone>. Because this application is used for publishing AIR applications, this node is used for any iPhone-specific configurations that you want to set. Adding an <InfoAdditions> node inside the <iPhone> node will allow you to add your custom iPhone settings.

The name of the key that sets the status bar color is UIStatusBarStyle. It has three valid values, which are all strings: UIStatusBarStyleDefault, which is the default gray bar, UIStatusBarSytleBlackTranslucent, which is a black status bar at 50% alpha, and UIStatusBarStyleBlackOpaque, which is an opaque black status bar.

After your application is published, you can rename the .ipa file to a .zip file and extract it. This will show you the contents of your application and Info.plist. You can open it in a text editor to see your additions.

Change the Status Bar Style

Change the Status Bar Style

1. Click File → Open.

   The Open dialog box appears.

2. Navigate to and click your application descriptor file.

3. Click Open.

4. If it does not already exists, add the iPhone node, such as `<iPhone></iPhone>`.

5. Add the info additions node, such as `<InfoAdditions></InfoAdditions>`.

6. Add a CDATA node, such as `<![CDATA[]]>`.

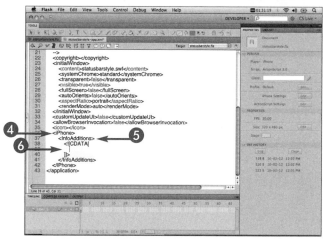

7 Add the status bar style key, such as `<key>UIStatusBarStyle</key>`.

8 Set the status bar style, such as `<string>UIStatusBarStyleBlackOpaque</string>`.

Note: `UIStatusBarSytleBlackTranslucent` and `UIStatusBarStyleDefault` *are also valid values.*

9 Click File → Save.

The status bar of your application will now appear according to the style that you set.

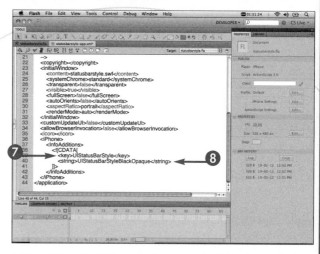

See the Status Bar Styles

You can check out what the different status bar styles look like using the iPhone Dev Center's Reference Library.

● This is the `UIStatusBarStyleDefault` style.

● This is the `UIStatusBarStyleBlackOpaque` style.

● This is the `UIStatusBarSytleBlackTranslucent` style.

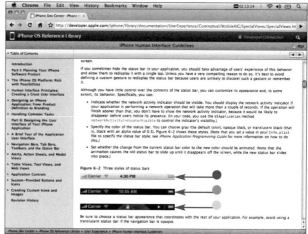

Chapter 4: Designing Your Application

Apply It

The `UIStatusBarStyle` key is just one of the many keys that you can add to the Info.plist file. You will explore more of them throughout the book. The Info.plist file contains a lot of other useful information about your application. Because the file is simply an XML text file, you can easily load the file and parse any data out of it that you may be interested in. Here is the syntax for reading the Info.plist file and creating an XML object with its contents:

```
var stream:FileStream = new FileStream();
stream.open( new File( "Info.plist" ), FileMode.READ );
var infoXML:XML = new XML( stream.readUTFBytes( stream.bytesAvailable ) );
```

Create Button States

One of the big differences between developing applications for the iPhone compared to a desktop computer is handling interaction. On a desktop, your application would make use of the mouse as a primary interaction input, whereas on a device, your fingers do most of the work.

The first thing you will probably realize is that you will not receive any MouseEvent.ROLL_OVER events because you cannot actually roll over any objects. Because of this, any buttons that may have rollover states for the desktop computer will not be shown. It is important, and good practice, to make sure that all your buttons have highlighted states, or mouse down states. This will give the users visual feedback that they actually touched on the button that they intended to touch. Creating buttons without proper states may lead the users to think that your application is broken or that the device is frozen.

There are two ways to make sure that your buttons have the proper states. The easiest way is to create a Button symbol in the Flash IDE and create a new state for your button in the appropriate frame on the Timeline. The second way is to listen for the MouseEvent.MOUSE_DOWN event on a MovieClip through code. This allows you to change the appearance of your MovieClip through code when a finger is touched down on the hit area of your button. If you use this method, you need to ensure that you change the state back when the user lifts his or her finger off the button as well.

Create Button States

1. Create a Button symbol.
2. Create a MovieClip.

Note: See Chapter 2, "Getting Started with Flash CS5," for more details on these steps.

3. Select the MovieClip on the Stage.
4. Give the button an instance name, such as mc_btn.

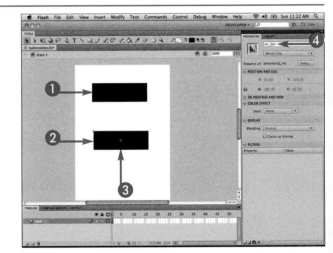

5. In the Timeline panel, click the New Layer button.

   A new layer is created.
6. Select the new layer.
7. Open the Actions panel.

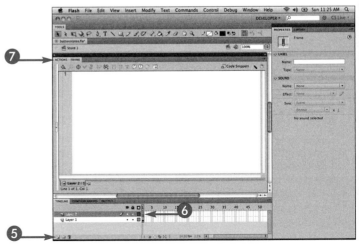

⑧ Add a mouse down listener to your MovieClip, such as `mc_btn.addEventListener( MouseEvent.MOUSE_DOWN, mouseDown );`.

⑨ Add a mouse up listener to your MovieClip, such as `mc_btn.addEventListener( MouseEvent.MOUSE_UP, mouseUp );`.

⑩ Create a `mouseDown` event handler.

⑪ Create a `mouseUp` event handler.

Note: For more details on creating event handlers, see Chapter 2.

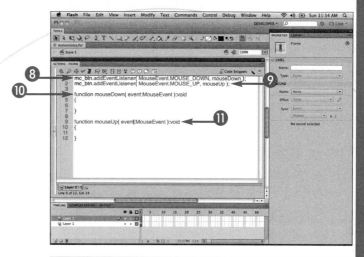

⑫ In your `mouseDown` function, create a new ColorTransform instance, such as `var trans:ColorTransform = new ColorTransform();`.

⑬ Set the color property, such as `trans.color = 0x000099;`.

⑭ Apply the ColorTransform, such as `mc_btn.transform.colorTransform = trans;`.

⑮ Repeat steps **12** to **14** in your `mouseUp` method and change the color, such as `trans.color = 0x00000;`.

⑯ Publish the file and click the buttons to see their states change.

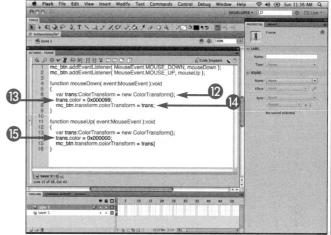

Extra

The other thing to keep in mind when creating buttons is size. On a desktop computer, you could have a 1 x 1 pixel button and still be able to click it with the mouse. On a touch screen, your finger is less precise, which makes it really hard to click small buttons. So it is important to make your buttons big enough in size that the users can register a proper touch. It is also a good idea to keep your buttons far enough apart so that the users do not select a button that they did not intend to. When designing buttons in your application, try to give them the same look and feel as the ones that Apple uses in its applications. Creating similar buttons will make your users familiar with your application right away. In this section, the example shows creating simple color changes for the different states; however, you can also create a state that shows a circular highlight representing where the finger pressed on the screen. This is a common practice and is often seen in Apple's applications.

Respond to Touch Events

For developing for platforms that can handle detecting a user's touch, Adobe has introduced the `flash.events.TouchEvent` class. This class is new and available on Flash Player 10.1, AIR 2.0, and the iPhone. Touch events can be thought of as mouse events for touch-enabled devices, in that they allow you to respond to basic touch interactions.

When a user interacts with the screen with a single or multiple fingers, multiple types of `TouchEvent`s will be fired that allow you to respond correctly to the user. You can listen for `TouchEvent`s on any `InteractiveObject`. Here is a typical sequence of events that will be fired during a single touch interaction: First, a `TouchEvent.TOUCH_BEGIN` event is fired when the user first presses his finger on the screen of the device. Second, a `TouchEvent.TOUCH_MOVE` is fired if the user drags his finger on the screen while the object is still pressed. Lastly, a `TouchEvent.TOUCH_END` event is fired when the user lifts his finger off the screen.

In order to respond to these events, you need to add a listener to an `InteractiveObject`. This can be a `Button`, a `MovieClip`, or even the root Stage of your application. In the example below, I add a few listeners to the main Timeline of a blank `.fla` to create a very simple paint program to help illustrate what I have talked about.

If you are going to develop your application to support other platforms other than iPhone, you can detect to see if `TouchEvent`s are supported on the device. `flash.ui.Multitouch.supportsTouchEvents` returns true if the device does support `TouchEvent`s and false if it does not. Planning ahead of time to support multiple platforms is never a bad thing; it will save you development time down the road.

Respond to Touch Events

1. In the Timeline panel, click the New Layer button.

   A new layer is created.

2. Select the new layer.

3. Open the Actions panel.

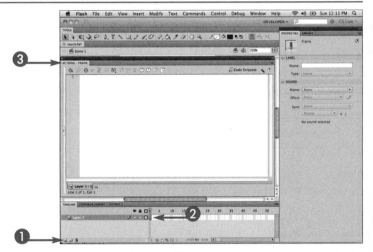

4. Set the multitouch input mode to touch, such as `Multitouch.inputMode = MultitouchInputMode.TOUCH_POINT;`.

5. Add a `TOUCH_BEGIN` listener to the Stage, such as `this.stage.addEventListener( TouchEvent.TOUCH_BEGIN, onTouchBegin );`.

6. Add a `TOUCH_MOVE` listener to the Stage, such as `this.stage.addEventListener( TouchEvent.TOUCH_MOVE, onTouchMove );`.

7. Add a `TOUCH_END` listener to the Stage, such as `this.stage.addEventListener( TouchEvent.TOUCH_END, onTouchEnd );`.

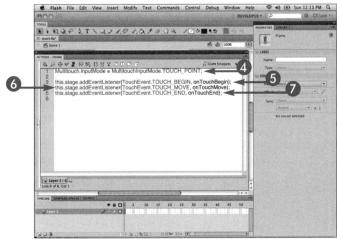

⑧ Create an `onTouchBegin` event handler for your listener.

⑨ Create an `onTouchMove` event handler for your listener.

⑩ Create an `onTouchEnd` event handler for your listener.

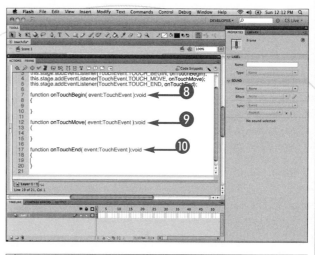

⑪ In the `onTouchBegin` method, set the line style, such as `this.graphics.lineStyle( 5, 0xFF0000 );`.

⑫ Move the drawing position to the touch location, such as `this.grapics.moveTo( event.stageX, event.stageY );`.

⑬ In the `onTouchMove` method, draw a line to the new touch location, such as `this.graphics.lineTo( event.stageX, event.stageY );`.

⑭ In the `onTouchEnd` method, clear the drawing, such as `this.graphics.clear();`.

⑮ Publish the file and install the application on your device.

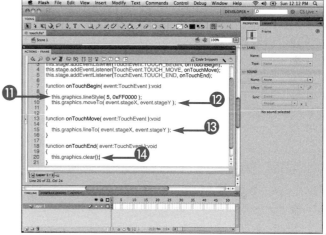

Extra

It is really important to plan ahead when creating your applications. Even if at the time of development, you do not plan on supporting any other platforms, you never know when Adobe will support a new platform that you did not expect during development. The less amount of time it takes to get your application working on a new platform, the better. If you do not plan to support multiple touches or gestures in your application, you may want to consider using `MouseEvent`s instead of `TouchEvent`s. This will allow your application to be compatible with more platforms than touch-enabled ones.

The sequence of events is similar to those used in the example in this section. `MouseEvent.MOUSE_DOWN` is fired when the user presses down on the mouse button, `MouseEvent.MOUSE_MOVE` is fired when the user moves the mouse with the mouse button still pressed, and `MouseEvent.MOUSE_UP` is fired when the user releases the mouse button. If you have previous experience developing Flash applications, this concept is not new to you, and you can apply that knowledge the same way you would when developing applications for the Web or desktop.

Track Multiple Touches

Tracking multiple touches is not much harder than tracking a single touch. The `TouchEvent.touchPointID` property is a unique ID for each unique touch that occurs. This ID is assigned when the `TouchEvent.TOUCH_BEGIN` event is fired and can be used to track unique touches on the screen.

For each new touch that is detected, the `touchPointID` is incremented. For example, if you were to place two fingers on the screen and drag them around, you would receive touch event objects with `touchPointID` values of 1 and 2. If you lifted those fingers and placed them back on the screen, you would get `touchPointID` values 3 and 4. This makes it a little harder to track specific fingers because `touchPointID` 2 does not necessarily mean finger 2. In order to track specific fingers, you will need to store which `touchPointID`s are currently being used in an `Array` or `Dictionary` object, as you will see in the example below. You place the ID in this object during the `TouchEvent.TOUCH_BEGIN` event and remove it in the `TouchEvent.TOUCH_END` event.

Each touch screen device has a different number of touches that it can detect at once. Currently, a Flash application on an iPhone and iPod touch can detect a maximum of five touches simultaneously. If you are planning on releasing your application on multiple platforms that support touch screens, you will want to plan for this number to change and be able to accommodate fewer or more touches. To do this, you can use the `Multitouch.maxTouchPoints` property to determine how many touches can be detected at the same time.

Track Multiple Touches

1. Set the input mode, such as `Multitouch.inputMode = MultitouchInputMode.TOUCH_POINT;`.

2. Create an array to hold a color for each finger, such as `var colors:Array = [ 0x731931, 0x401323, 0x262226, 0x54594C, 0x888C65];`.

3. Create a `Dictionary` instance to hold sprite references, such as `var sprites:Dictionary = new Dictionary();`.

4. Create a touch counter, such as `var touchCount:int = 0;`.

5. Add listeners and event handlers for the `TOUCH_BEGIN`, `TOUCH_MOVE`, and `TOUCH_END` events.

6. Create a new `Sprite` instance and add it to the Stage, such as `var mc:Sprite = new Sprite(); addChild( mc );`.

7. Set the sprite's `lineStyle`, such as `mc.graphics.lineStyle( 5, colors[ touchCount ] );`.

8. Set the initial drawing location to the touch position, such as `mc.graphics.moveTo( event.stageX, event.stageY );`.

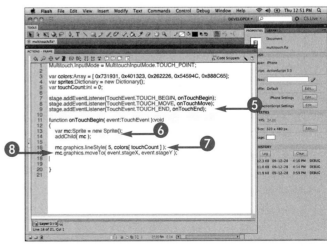

⑨ Increment the touch count, such as `touchCount++;`.

⑩ Store a reference to the sprite, such as `sprites[ event.touchPointID ] = mc;`.

⑪ Get the reference to the sprite, such as `var mc:Sprite = sprites[ event.touchPointID ] as Sprite;`.

⑫ Draw a line to the new touch position, such as `mc.graphics.lineTo( event.stageX, event.stageY );`.

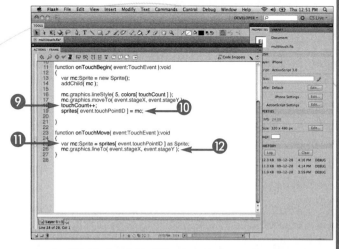

⑬ Get the reference to the sprite, such as `var mc:Sprite = sprites[ event.touchPointID ] as Sprite;`.

⑭ Remove the sprite from the Stage, such as `this.removeChild( mc );`.

⑮ Decrement the touch count, such as `touchCount--;`.

⑯ Delete the reference to your sprite, such as `delete sprites[ event.touchPointID ];`.

⑰ Publish the file and install it on your device. Draw with multiple fingers on the screen.

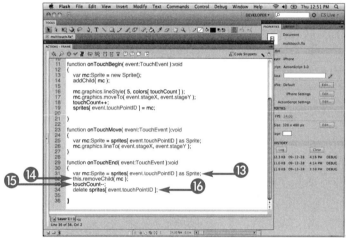

Extra

If you would like to play around with a different color scheme, you can easily find new ones in the Kuler panel. You can access this panel by clicking Window → Extensions> → Kuler in Flash Professional CS5. On this panel, you can browse different color schemes that other users have created. When you have found a scheme that you like, you can add the colors to the Swatches panel. If you do not find one that suits you, you can always edit an existing one or create a new one.

Respond to Zoom Events

One of the main user interactions for multitouch devices is the pinch and zoom gesture. It is most commonly used to scale objects up and down. You have probably used this gesture yourself a number of times to zoom in and out of Web sites in Safari or email on your device. If you are new to the platform, the zoom gesture, sometimes referred to just as *pinch,* is achieved by placing two fingers on the screen and moving them apart to zoom in and moving them closer to zoom out. It would be possible to achieve this effect by tracking multiple touch points, but that is harder than it sounds. Luckily, Adobe has provided the `TransformGestureEvent.GESTURE_ZOOM` event, which will take care of all the hard work.

In order to respond to this gesture, you simply add an event listener to any `InteractiveObject`, just as you have done in previous examples. When the event is fired, the event handler method will be called with a `TransformGestureEvent` object as a single argument.

This event contains all the information you need in order to scale the object in relation to a gesture. In particular, the two properties of interest are `scaleX` and `scaleY`. Your first thought may be to simply set the scale values that are returned from the event to your object, but that will give you an undesired result. The reason for this is that the scale properties that are returned are values based on the previous gesture event — not its current scale value. In order to calculate your object's new scale properties, you simply multiply the object's current scale by the event's scale values.

Respond to Zoom Events

① Create a `MovieClip`.

Note: *See Chapter 2 for more details on this step.*

② Give it an instance name, such as `square_mc`.

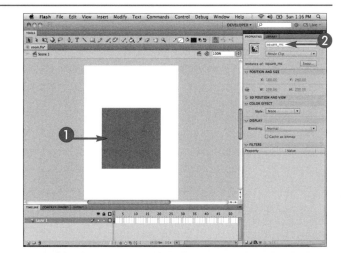

③ In the Timeline panel, click the New Layer button.

A new layer is created.

④ Select the new layer.

⑤ Open the Actions panel.

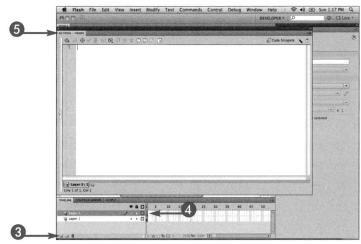

6 Set the input mode, such as `Multitouch.inputMode = MultitouchInputMode.GESTURE;`.

7 Add a listener for the zoom event, such as `square_mc.addEventListener( TransformGestureEvent.GESTURE_ZOOM, onZoom );`.

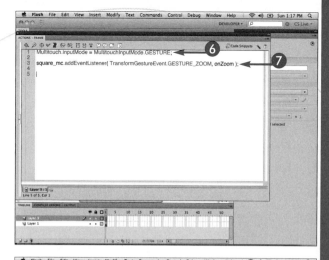

8 Create an event handler for your zoom listener.

Note: See Chapter 2 for more details on creating event handlers.

9 Set the `scaleX` of the square based on the gesture, such as `square_mc.scaleX *= event.scaleX;`.

10 Set the `scaleY` of the square based on the gesture, such as `square_mc.scaleY *= event.scaleY;`.

11 Publish the file and install it on your device.

12 Place two fingers close together on the square and move them apart.

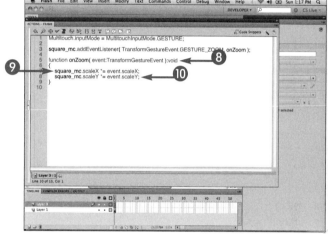

Extra

A tip for creating smooth-looking zoom gestures is to make sure that your registration point is at the center of your object instead of the default top left, or 0,0. This will ensure that your object is scaled from its center and will be consistent with all other objects that react to this gesture in other applications on your device. There are a couple of different ways that you can set the registration point. The easiest is way is to set it in the Convert to Symbol dialog box by selecting the center dot as the registration point. The other way is through code and will require you to have your object in a parent `DisplayObjectContainer`:

```
child.x = -child.width;
child.y = -child.height;
```

This will center the child object to its parent's registration point, assuming that it is at 0,0.

Respond to Rotate Events

Another commonly used multitouch interaction, aside from the pinch and zoom gesture, is the two-finger rotate gesture. To rotate the object, place one finger on it and rotate a second finger around the first. It is important to note that only one finger needs to be on the object that you are rotating. The second finger can be outside of the hit area of the object, but it must remain on the screen at all times.

In order to detect for this gesture, you can listen for the `TransformGestureEvent.GESTURE_ROTATE` event on any `InteractiveObject` on the Stage. If a rotate gesture is detected, this event will be fired, and your event handler will be called with a `TransformGestureEvent` object as its argument. This object contains all the information that you need in order to rotate your object. The `TransfromGestureEvent.rotation` property contains the rotation of the object since the previous rotation event. In order to have this affect your object, you add the event's rotation value to the current rotation value of your object.

It is a good idea for all the objects that respond to a rotation gesture to have their registration be at their center. This will ensure that your fingers are always on the object as you rotate it. If you have your object's registration point at the default top left, or 0,0, you could rotate the object off the screen, which could produce some undesirable results.

Respond to Rotate Events

① Create a `MovieClip`.

Note: See Chapter 2 for more details on this step.

② Give it an instance name, such as `square_mc`.

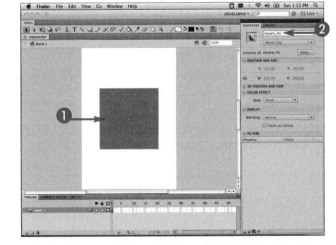

③ In the Timeline panel, click the New Layer button.

A new layer is created.

④ Select the new layer.

⑤ Open the Actions panel.

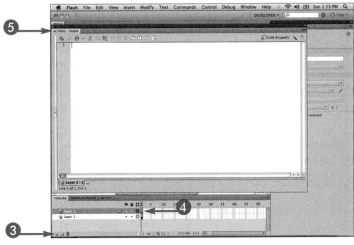

6. Set the input mode, such as `Multitouch.inputMode = MultitouchInputMode.GESTURE;`.

7. Add a listener for the rotate gesture, such as `square_mc.addEventListener ( TransformGestureEvent.GESTURE_ROTATE, onRotate );`.

8. Create an event handler for your rotate listener.

Note: *See Chapter 2 for more details on this step.*

9. Set the new rotation, such as `square_mc.rotation += event.rotation;`.

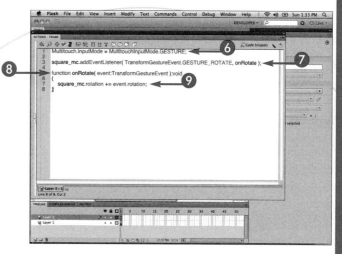

10. Publish and install the application on your device.

11. Place two fingers on the square and rotate one around the other.

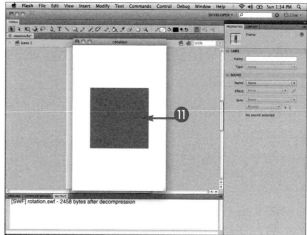

Apply It

If you are feeling adventurous and want to try and program this gesture yourself, you can do so by tracking two fingers. One way to determine the new rotation of your object is to find the angle between two lines. To do this, you will need to store the previous location of the two touches. Those two points will form line 1, and the current touch points will form line 2. Here is a method that calculates the angle in degrees from two lines:

```
function angleBetweenLines( line1Start:Point, line1End:Point, line2Start:Point,
  line2End:Point):Number{
var a:Number = line1End.x - line1Start.x;
var b:Number = line1End.y - line1Start.y;
var c:Number = line2End.x - line2Start.x;
var d:Number = line2End.y - line2Start.y;
var degs:Number = Math.cos(((a*c) + (b*d)) / ((Math.sqrt(a*a + b*b)) * (Math.sqrt(c*c +
  d*d))));
return degs * ( 180 / Math.PI );
}
```

Respond to Pan Events

Because of the limited screen real estate, there will be times when your content will be bigger than the viewing area. When this happens, you will want to give the users the ability to scroll or pan the content so that they can see any hidden content. This can also happen when you zoom in on objects, and objects scale up past the bounds of the screen. For example, consider a photo gallery application. When you initially display the image, you would show it at a size that fills the screen. But you may allow the user to zoom into specific areas of the image using the `TransformGestureEvent.GESTURE_ZOOM` event. The user scales the image up twice its original size, and now the part of the image that she is interested in is off the screen. You will want to let her pan the image around in order to see any areas that were off-screen.

To detect for a pan gesture, you can listen for the `TransformGestureEvent.GESTURE_PAN` event on any `InteractiveObject`. This event is fired when it is detected that the user has placed two fingers on the object and is dragging them around the screen. Only one finger must be on the targeted object, and the other controls the direction in which to pan the object. The `TransformGestureEvent` object that gets passed to the event handler contains two properties that you will use to move your object. `TransformGestureEvent.offsetX` and `TransformGestureEvent.offsetY` give the difference in position since the last pan event. To move the targeted object, add the current x and y values of your object to these values.

Respond to Pan Events

Set Panning on an Image

1. Import an image onto the Stage.

Note: See Chapter 6, "Working with Images," for more details.

2. Convert the image to a `MovieClip`.

Note: See Chapter 2 for more details on this topic.

3. Give the `MovieClip` an instance name, such as `image_mc`.

4. In the Timeline panel, click the New Layer button.

   A new layer is created.

5. Select the new layer.

6. Open the Actions panel.

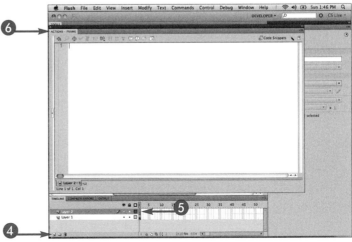

7 Set the input mode, such as `Multitouch.inputMode = MultitouchInputMode.GESTURE;`.

8 Add a listener for the pan gesture, such as `image_mc.addEventListener( TransformGestureEvent.GESTURE_PAN, onPan );`.

9 Create an event handler for your pan listener.

Note: See Chapter 2 for more details on this step.

10 Set the new x position based on the gesture, such as `image_mc.x += event.offsetX;`.

11 Set the new y position based on the gesture, such as `image_mc.y += event.offsetY;`.

Test the Application

12 Publish and install the application on your device.

13 Place two fingers on the image and move them around the screen to pan the image.

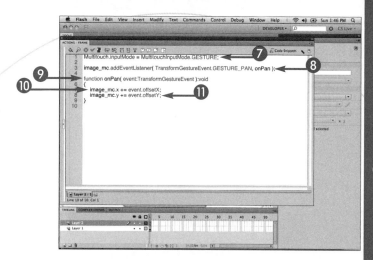

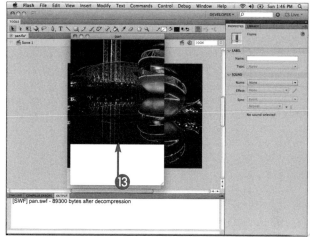

Apply It

There are other ways to pan objects than using the `TransformGestureEvent.GESTURE_PAN` event. For example, you could use `TouchEvent.TOUCH_MOVE` to drag the object around if you wanted to, as in the following:

```
Multitouch.inputMode = MultitouchInputMode.TOUCH_POINT;
this.stage.addEventListener(TouchEvent.TOUCH_MOVE, onTouchMove);
function onTouchMove( event:TouchEvent ):void{
image_mc.x = event.stageX;
image_mc.y = event.stageY;
}
```

However, you cannot detect for gestures and touches at the same time in your application. So if want to detect for any other gestures and want the ability to pan and drag, you will need to use the `TransformGestureEvent.GESTURE_PAN` **event**.

Respond to Swipe Events

The swipe is a relatively simple gesture. It can be initiated by swiping your finger either vertically or horizontally in a straight line with a single finger. Apple uses this gesture in a lot of its applications and in various ways. The Photo application uses it to scroll through the photos, whereas the Mail application uses it to show the Delete button on an email item.

You can listen for a swipe gesture on any `InteractiveObject` by adding listening for its `TransformGestureEvent.GESTURE_SWIPE` event. When a swipe gesture is detected, a `TransformGestureEvent` object is passed as an argument to your event handler. The `offsetX` property will return either 1 for a swipe right and -1 for a swipe left. The `offsetY` property will return either 1 for a swipe down and -1 for a swipe up.

As soon as you know which way the swipe occurred, you can adjust your content accordingly.

The example below shows importing two images to the Stage and moving them left or right depending on the direction of the swipe. When you first load the application, you are going to place the first image on the Stage and have it take up the entire screen area. Because it is the first image, you are only going to allow for left swipes. If you swiped the content right, there would not be an image to move into its place. When you swipe left, you are going to animate the current image off-screen to the left and animate the second image on screen from the right. When the second image is shown, you are going to check for right swipes. When a right swipe is detected, you are going to animate the second image off-screen to the right and animate the first image in from the left.

Respond to Swipe Events

Set Swiping for Two Images

1. Import an image onto the Stage.

Note: See Chapter 6 for more details.

2. Convert the image to a `MovieClip`.

Note: See Chapter 2 for more details.

3. Place the image at 0,0.

4. Give it an instance name, such as `image1_mc`.

5. Import a second image onto the Stage.

Note: See Chapter 6 for more details.

6. Convert the image to a `MovieClip`.

Note: See Chapter 2 for more details.

7. Place the image at 320,0.

8. Give it an instance name, such as `image2_mc`.

⑨ Open the Actions panel.

⑩ Set the input mode, such as `Multitouch.inputMode = MultitouchInputMode.GESTURE;`.

⑪ Add a swipe gesture listener to both images.

⑫ Create an event handler, such as `onSwipe`.

⑬ Check to see which image detected the swipe gesture.

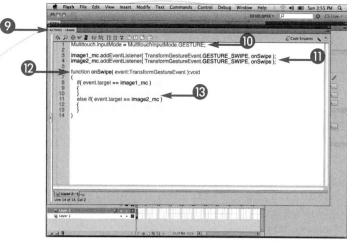

⑭ Check to see if it was a left swipe.

⑮ If so, animate the first image to -320 x and the second image to 0 x.

⑯ Check to see if it was a right swipe.

⑰ If so, animate the first image to 0 x and the second image to 320 x.

Test the Application

⑱ Publish and install the application on your device.

⑲ Swipe the images left and then swipe them right.

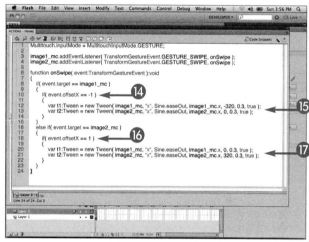

Apply It

The way shown in this section is the easiest way to detect for swipes, but you can also detect touches independently and determine which direction a swipe is going. By doing this, you have more control over how fast and how far the swipe has to go before it is classified as a swipe:

```
image1_mc.addEventListener( TouchEvent.TOUCH_BEGIN, onBegin );
image1_mc.addEventListener( TouchEvent.TOUCH_MOVE, onMove );
var startX:Number;
var startY:Number;
function onBegin( event:TouchEvent ):void{
    startX = event.stageX;
    startY = event.stageY;
}
function onMove( event:TouchEvent ):void{
if( startX > event.stageX ){
        // left swipe detected
    }else{
        //right swipe detected
    }
}
```

Listen for Accelerometer Events

A popular interaction with the iPhone and iPod touch, especially among games, is using the accelerometer. Driving games use the accelerometer, allowing the users to rotate their devices back and forth in order to steer their vehicle. Also, shaking the device has become a popular way to incorporate Easter eggs into your game or application. Apple also uses the accelerometer to detect for a shake to initiate an Undo command in many of its applications.

The new `Accelerometer` class lets you receive acceleration data from the onboard accelerometer chip of the device. As the device moves, you will receive linear acceleration data along the x, y, and z axes. To receive this data, listen for the `AccelerometerEvent.UPDATE` event on the `Accelerometer` object. An `AccelerometerEvent` object will be passed to your event handler, which contains `accelerationX`, `accelerationY`, and `accelerationZ` properties, a value for each axis.

The `accelerationX` property represents the acceleration measured in Gs along the x-axis. This axis runs from the left to the right of the device when it is in its upright position. The `accelerationY` property represents the acceleration measured in Gs along the y-axis. This axis runs from the bottom of the phone to the top. In the case of an iPhone, this axis runs from the home button to the earpiece. The `accelerationZ` property represents the acceleration measured in Gs along the z-axis. This axis runs perpendicular to the face of the phone, and the value will be positive as it moves closer to you.

The example below shows moving a ball around the screen as the device is tilted and moved around.

Listen for Accelerometer Events

Create a Ball That Moves with the Device

1. Create a circle `MovieClip`.
2. Give it an instance name, such as `circle_mc`.

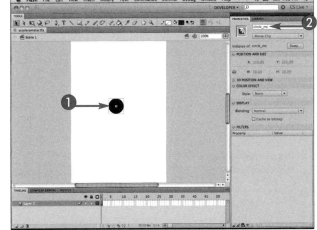

3. Create a `velocityX` variable, such as `var velocityX:Number = 0;`.
4. Create a `velocityY` variable, such as `var velocityY:Number = 0;`.
5. Create an `Accelerometer` variable, such as `var am:Accelerometer = new Accelerometer();`.
6. Set the update interval of the accelerometer.
7. Add an update listener to the accelerometer.
8. Create an event handler, such as `onUpdate`.

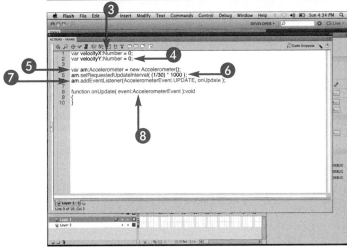

9. Create a Number variable, such as var velocityMultiplier: Number = 0.5;.

10. Apply the accelerationX value to calculate the velocity of x.

11. Apply the accelerationY value to calculate the velocity of y.

12. Set the x property of the circle instance, such as circle_mc.x += velocityX.

13. Set the y property of the circle instance, such as circle_mc.y -= velocityY.

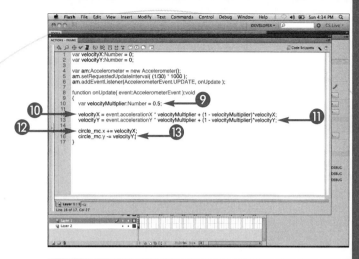

Test the Application

14. Publish and install the application on your device.

15. Move the device around to move the circle on the screen.

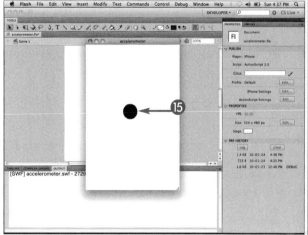

Apply It

One thing to keep in mind when trying to animate an object with values from the accelerometer is that you are not guaranteed to get values at a regular interval. So if you are trying to create nice, smooth, time-based animations based on that data, you will need to store the time between each accelerometer update and factor that into your equation.

Another option is to animate the object on a Timer or EnterFrame event and only update the velocities in the accelerometer update handler. To do so, remove the following code from the onUpdate method:

```
circle_mc.x += velocityX;
circle_mc.y -= velocityY;
```

And add the following code to the example:

```
stage.addEventListener( Event.ENTER_FRAME, onFrame );
function onFrame( event:Event ):void{
    circle_mc.x += velocityX;
    circle_mc.y -= velocityY;
}
```

Determine If the Accelerometer Is Available

When developing an application that takes advantage of the accelerometer, you are going to need to place the application on your device in order to test it. This can slow down development if every time you make changes to your application, you need to test it on the device. This is extremely inefficient, especially if you are not testing the portion of your application that uses the accelerometer. To get around this issue, a good practice is to detect to see if the accelerometer is available and give yourself another way of interacting with your application if it is not.

For example, if you need to detect for a shake motion, you could place a button on the screen that when pressed would simulate a shake. Doing this allows you to test on your computer, and when you are satisfied that everything is working, you can put your application on the device to test your accelerometer code.

The `Accelerometer` class has a static property on it named `isSupported` to detect if it is available. It is a read-only property that if set to true it is available; otherwise, it is false.

The example below shows comparing the current acceleration data to the previous to see if there has been a big enough shake. If a shake has occurred, you are going to play a sound. To detect for the shake, compare the previous acceleration data of all three axes from the current data. If there was enough of a difference in acceleration in any direction, you can determine that the user has shaken the device.

Determine If the Accelerometer Is Available

Respond to a Shake with a Sound

① Import a sound effect audio file.

② Give the sound a class name, such as `Shake`.

Note: *For more details on importing sounds, check out Chapter 7, "Working with Sound."*

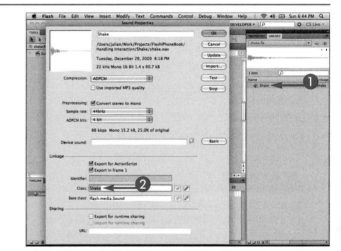

③ Open the Actions panel.

④ Create a `Sound` variable for the shake sound, such as `var sound:Sound = new Shake();`.

⑤ Check to see if the accelerometer is available.

⑥ Create `Number` variables for the previous x, y, and z positions, such as `lastX`, `lastY`, and `lastZ`.

⑦ Create a new `Accelerometer` variable, such as `var am:Accelerometer = new Accelerometer();`.

⑧ Add an update listener to the `Accelerometer`.

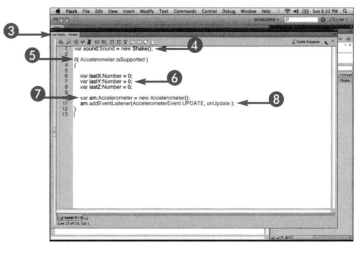

⑨ Create a `Sprite` and draw a rectangle in it.

Note: See Chapter 2.

⑩ Add the sprite to the Stage.

⑪ Add a listener for the mouse click.

⑫ Create an event handler, such as `onClick`, and call the `shake` method inside it.

⑬ Create a `shake` function.

⑭ Play the sound, such as `sound.play();`.

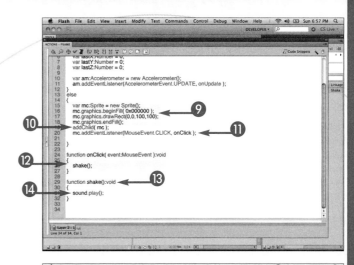

⑮ Create an event handler, such as `onUpdate`.

⑯ Create a `Number` variable, such as `var threshold:Number = 0.9;`.

⑰ Calculate the difference between the current data and the previous data.

⑱ Store the current acceleration for the next update.

⑲ Determine if there was a big enough difference in any of the axes and call the `shake` method.

Test the Application

⑳ Publish the application and run it on your device and the desktop to see the difference.

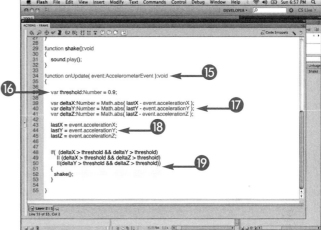

Apply It

There is also an `Accelerometer.muted` property that you can use to detect if the accelerometer is available. This is used to determine if the user has denied access to the accelerometer. Currently, the iPhone and iPod do not have an option that allows the user to disable the accelerometer, but one may exist on other platforms. So if you are targeting multiple platforms, you may want to check this property as well for determining if the accelerometer is available. If you are just building iPhone applications, you do not have to worry about it. Here is how to use this property:

```
if( Accelerometer.isSupported || !Accelerometer.muted )
{
//Accelerometer is available.
}
```

Determine Device Orientation

Determining the orientation of the device can be easily achieved by reading the acceleration data from the accelerometer. You can detect to see if the device is positioned on any one of its six sides. This can be used in order to rotate content so that it matches the rotation of the device. It can also be used to detect user input. For example, say that you wanted the users to flip an object in a game. You could have the users actually flip their devices in order to flip the object in the game. Keep in mind, though, that the more extreme the gesture, the bigger the chance the users can have their devices flying out of their hands. And nobody will be doing that gesture more than you during development.

Once you are familiar with reading the acceleration data from the accelerometer, determining the orientation is pretty simple. If the accelerationX > 0.5, then the device is laying on its right side, the side opposite the volume controls. If the accelerationX < -0.5, the device is laying on its left side, the side with the volume controls. If the accelerationY > 0.5, the device is laying on its top side, the side with the headphone jack and lock button. If the accelerationY < -0.5, the device is standing up, with the side with the connector jack pointing to the floor. If the accelerationZ > 0.5, the device is lying with its screen down. If the accelerationZ < -0.5, the device is lying with its back down.

Determine Device Orientation

Create a TextField That Displays the Device Orientation

1. Click the Text tool.
2. Create a TextField on the Stage.
3. Click here and select Dynamic Text as the TextField's type.
4. Give the text field an instance name, such as debug_txt.

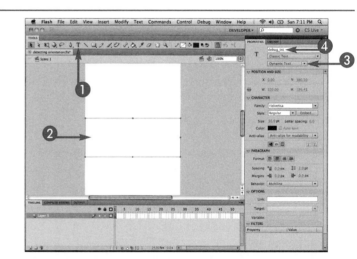

5. Create a new Accelerometer variable, such as var am:Accelerometer = new Accelerometer();.
6. Add an update event listener to the accelerometer variable.
7. Create an event handler, such as onUpdate.

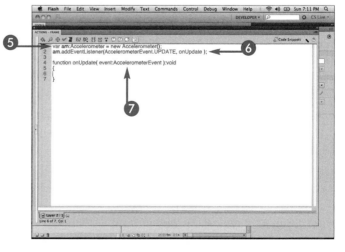

8 Check to see if the `accelerationX` is greater than 0.5.

9 Set the text field, such as `debug_txt.text = "Right Side";`.

10 Check to see if the `accelerationX` is less than -0.5.

11 Set the text field, such as `debug_txt.text = "Left Side";`.

12 Check to see if the `accelerationY` is greater than 0.5.

13 Set the text field, such as `debug_txt.text = "Upside Down";`.

14 Check to see if the `accelerationY` is less than -0.5.

15 Set the text field, such as `debug_txt.text = "Standing Up";`.

16 Check to see if the `accelerationZ` is greater than 0.5.

17 Set the text field, such as `debug_txt.text = "Face Down";`.

18 Check to see if the `accelerationZ` is less than -0.5.

19 Set the text field, such as `debug_txt.text = "Face Up";`.

Test the Application

20 Publish and install the application on your device. Change the orientation.

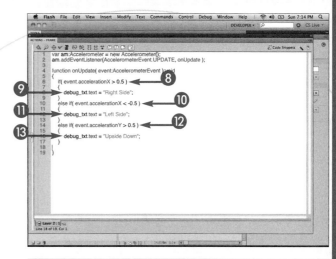

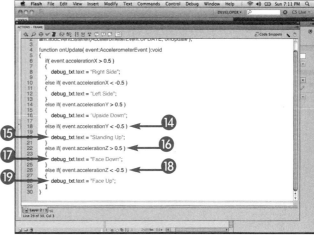

Extra

After you have determined the orientation of your device, you will want to rotate your content for your user. If your device is lying on its left side, you will want to rotate your content 90 degrees. If your device is lying on its right side, rotate your content -90 or 270 degrees. If your device is upside down or lying on its top, rotate your content 180 degrees. For the other three sides, the content's rotation should be set to 0 degrees. One thing to keep in mind when rotating content this way is that you will have to reposition it as well. You could also have separate views for each rotation and swap them as the device rotates. The advantage to this is that you can potentially show different sized graphics when the device is rotated to either its left or right side to maximize the change in screen real estate. There is no right way to do it, and you will have to decide which way works best for you on a case-by-case basis.

Detect Which Way Is Up

When you are developing applications that allow the user to rotate the device, it will be important for you to make sure that your content is rotated and displayed properly. You could implement something similar to the example in the section "Determine Device Orientation," or you could detect which side is pointing up. Detecting which side is pointing up allows you to rotate your content to a more precise angle, instead of just every 45 degrees.

First, set up the Accelerometer so that you can receive the update events. If you are unsure of how to do this, have a look at the example in the section "Listen for Accelerometer Events." After you have an event handler created and are receiving acceleration data, you can use the Math.atan2 method in order to calculate the angle in radians. The Math.atan2 method takes two arguments, y and x. It is important to note that the y property is always the first argument, which is usually different in any other method that accepts x and y properties as arguments. To calculate the angle, you pass in the accelerationY value and the negative value of accelerationX. Because a DisplayObjects.rotation property is expecting an angle in degrees, you will need to convert the radians to degrees in order to set the rotation of your content. The following equation shows you how to do the conversion:

```
var degrees:Number = radians * ( 180 /
  Math.PI );
```

After you have the angle in degrees, you can set it to the rotation property of your object. Below is a very simple example that rotates an image of an arrow to which way points up.

Detect Which Way Is Up

Create an Arrow to Point Up

1. Import an arrow graphic and convert it to a MovieClip.

Note: See Chapter 2 for more details on this topic.

2. Give it an instance name, such as arrow_mc.

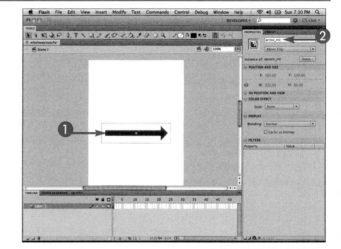

3. Open the Actions panel.

4. Create a new Accelerometer variable, such as var am:Accelerometer = new Accelerometer();.

5. Add an update event listener to the accelerometer variable.

6. Create an event handler, such as onUpdate.

7 Calculate the radians based on the acceleration values.

8 Convert the radians to degrees.

9 Set the rotation, such as `arrow_mc.rotation = degrees;`.

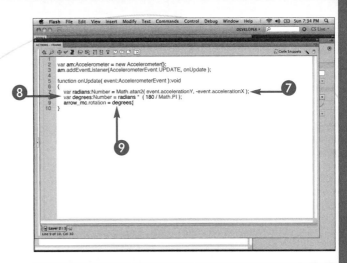

Test the Application

10 Publish and install the application on your device.

11 Rotate the device around to rotate the arrow.

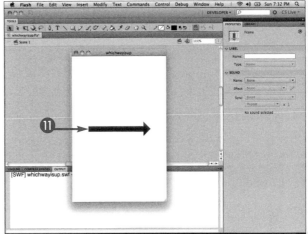

Apply It

Changing the example so that the arrow is pointing to the side that is down is relatively easy. You simply need to change the values that you pass into the `Math.atan2` function to be negative `accelerationY` and positive `accelerationX`. This can be used to make sure that objects always fall the direction that gravity is pulling them. Here is what the function looks like with these changes:

```
var radians:Number = Math.atan2( -event.accelerationY, event.accelerationX );
```

Filter Accelerometer Data

If you have experimented with the accelerometer, you have probably noticed that the data can sometimes contain noise. Depending on how you are visualizing the data, this noise may cause your graphics to jump around quite a bit. In order to reduce the noise, you will want to smooth out the values from the `Accelerometer` by filtering out unwanted values. Smoothing data sets is used to capture patterns in the data while removing any noise. This technique is often used in analyzing images and sound waves.

The `accelerationZ` property in the `AccelerometerEvent` object represents gravity. To better understand this, place your device on a flat surface, like a table. You should see that the `accelerationX` and `accelerationY` properties are approximately 0, and the `accelerationZ` property should be approximately -1.

In order to remove the effects of gravity, you can use a *high-pass filter*, which reduces the amplitude of the cutoff frequency. This filter reduces some of the noise reported to you by the accelerometer and gives you smoother results between updates.

The `dt` variable in the example below represents a time interval, and the `RC` variable represents a time constant. These two variables are used to calculate the `filterConstant` variable. A large `filterConstant` variable suggests that the data will decay very slowly over time and will also be strongly influenced by small changes in the accelerometer. A small `filterContstant` variable suggests that the data will decay quickly and will require large changes in the accelerometer in order to change the output. Changing the `rate` and `freq` variables in this example gives you more control over how you would like your acceleration data to be filtered.

Filter Accelerometer Data

Using a High-Pass Filter

1. Create a `Number` variable, such as `var rate:Number = 60;`.
2. Create a second `Number` variable, such as `var freq:Number = 5;`.
3. Create a third `Number` variable, such as `var dt:Number = 1.0/rate;`.
4. Create a fourth `Number` variable, such as `var RC:Number = 1.0/freq;`.

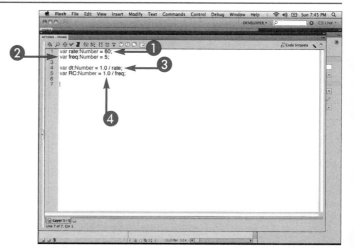

5. Create `Number` variables to store the acceleration data.
6. Create `Number` variables to store the filtered acceleration values.

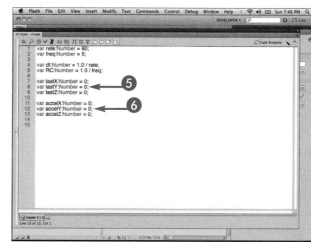

⑦ Create a `filter` function.

⑧ Create a `Number` variable, such as `var filterConstant :Number= RC / (dt + RC);`.

⑨ Filter the data for all three axes.

⑩ Store the current acceleration values for the next update.

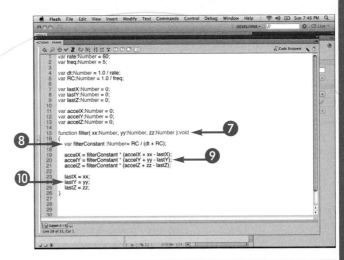

⑪ Create a new `Accelerometer` variable, such as `var am:Accelerometer = new Accelerometer();`.

⑫ Add an update event listener to the accelerometer variable.

⑬ Create an event handler, such as `onUpdate`.

⑭ Call the `filter` function.

Publish the Application

⑮ Publish and install the application on your device.

Apply It

You can also experiment with using a low-pass filter instead of a high-pass one. A low-pass filter reduces the amplitude of frequencies higher than the frequency cutoff and isolates the effects of gravity. To use this type of filter, you can simply replace the code in the `filter` method in the example shown here with the following:

```
var filterConstant:Number = dt / (dt + RC);
accelX = xx * filterConstant + accelX * (1.0 - filterConstant);
accelY = yy * filterConstant + accelY * (1.0 - filterConstant);
accelZ = zz * filterConstant + accelZ * (1.0 - filterConstant);
```

I encourage you to experiment with both filters and see which one works best for your specific implementation.

Prepare Your Images

Preparing your images before they are used in Flash can go a long way in the performance of your application. There are certain things that you will want to avoid and take into consideration when creating your images.

If you do most of your Flash development targeting the Web, you are used to trying to make your images as optimized as possible to reduce the file size. This usually means creating JPEG images and adjusting their compression settings. However, on the iPhone, PNG is the recommended image format to use. PNG images have an optimized drawing path to the screen and should be used for all your images, even if they do not contain any alpha.

The iPhone also has a maximum image size of 1024 x 1024. If, for some reason, your images are bigger than that, you will need to slice them up into multiple images. If they are bigger than the maximum dimensions, they will not appear on-screen.

It is also important to create your images at the size that they are intended to be shown at. Never create them larger than needed and scale them down to size in Flash. Scaling your images is a very performance-intensive operation and should be avoided whenever possible. Also, make sure that there is not any unnecessary alpha in your images. If your images require a transparent background, crop them so that there is the least amount of alpha in the image as possible.

Following these simple guidelines will help you get the best performance out of your images in your applications.

Prepare Your Images

1. Open Photoshop, or some other image-editing software.
2. Click File → Open and open the image to prepare.
3. Click the Crop tool.
4. Crop the image to reduce unwanted alpha.

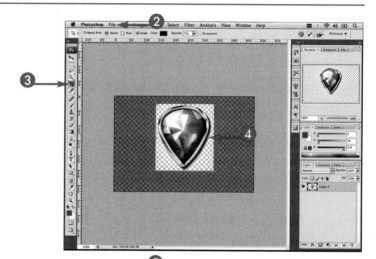

5. Click File.
6. Click Save As.

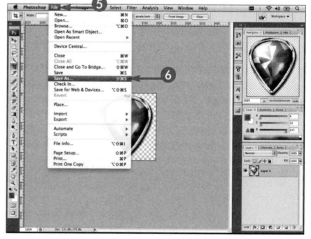

The Save As dialog box appears.

7 Click here and select PNG.

8 Name your file, such as gem.png.

9 Click Save.

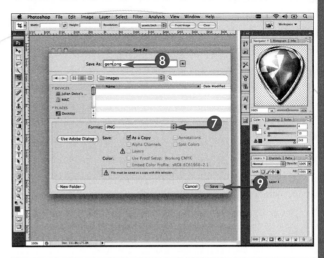

The PNG Options dialog box appears.

10 Click None.

11 Click OK.

The image is saved and is prepared to use in your application.

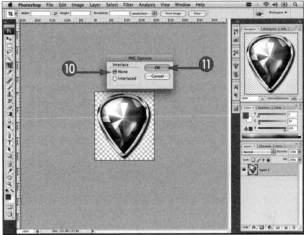

Extra

The iPhone video memory uses a nonstandard byte ordering of BGR (blue, green, red) instead of the standard RGB, whereas the PNG image format uses the standard byte ordering system of RGB. When you are developing applications with the iPhone SDK, during the build process, Xcode processes the PNG images in your application and swaps its bytes to use the BGR format using the pngcrush application. Your original file is left intact, and the optimized image is copied to your application bundle. This gives you performance increases because iOS does not have to process the image at runtime. This is why PNG images are the recommended image format. Images in any other format must be processed at runtime before being sent to the screen, which will cause a decrease in performance if you are trying to display a lot of images at the same time. Currently, Flash CS5 iPhone applications do not take advantage of this; however, I am confident that this will be available in future updates. When this feature does become available, you will be ready to take advantage of it right away and should see a performance increase in your applications.

Import Images

One way to include images in your application is to import them to the Library inside your Flash file. There are several advantages to this, the biggest one being that you are able to lay out your images accurately inside your file on the Stage.

When you import images to the Library, Flash compiles them into your .swf file. When you publish your iPhone application, your entire .swf file is compiled into the binary of your application. The more images in your Library, the bigger file size your application binary will have. When your application launches, your entire application is loaded into memory on the device. Therefore, the bigger your application binary, the longer the application will take to launch, and the more memory it will take up, even if you are not using the assets.

The advantage of this is that your images will be somewhat cached and you will not have to load them into memory every time, but this also means that you cannot remove them from memory. This is okay for images that you will use often, such as character animations in a game.

There are two options for importing images. You can import images to the Stage, or you can import them directly to the Library.

If you have an image sequence that you want to import, and if it is named properly, Flash will automatically detect that it is a sequence and import each image onto a separate frame on the Timeline. Images simply need to be named sequentially with a number suffix for Flash to recognize this. Simply select the first image in the sequence when importing them.

Import Images

Import an Image to the Library

1. Click File.
2. Click Import.
3. Click Import to Library.

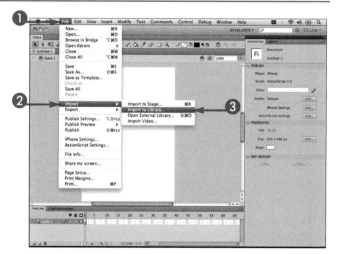

The Import to Library dialog box appears.

4. Click here and select All Image Formats.
5. Navigate to and select the image file that you want to import.
6. Click Open.

The image is imported to the Library.

7. Click Window → Library to open the Library panel.

• You can see your imported image here.

Note: For each image that you import, a graphic symbol is created for it in the Library. You can delete this symbol if you like, and just use the bitmap.

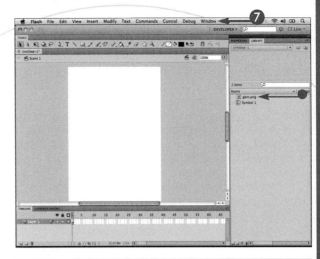

Place the Image on the Stage

8. Select your image in the Library.
9. Drag it onto the Stage.

Your image appears on the Stage.

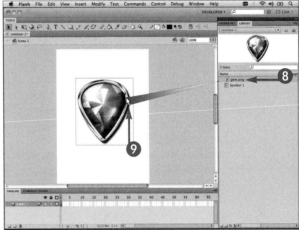

Extra

After you have imported your images to the Library, you can edit them in an image-editing program, such as Photoshop. This allows you to make edits to your image and have them automatically updated in Flash. To edit an image, right-click it in the Library and select Edit with Adobe Photoshop CS5. If you do not have Photoshop CS5 installed on your machine, you can select the Edit With option instead, and it will prompt you to select your image-editing application. After your image-editing program has launched, your image will be opened so that you can begin editing it. After your edits are complete, save the image. When you go back to Flash, you will see your image updated in the Library. It will also be updated anywhere on the Stage where the image is placed.

Alternatively, you can simply overwrite the file on your hard drive with a new version and select Update from the right-click menu of the image. This allows you to update an image in the event that you are not the one editing the images.

Display Images

After you have added images to your Flash file, you will want to display them. The easiest way is to simply drag them from the Library onto the Timeline. Adding images to the Timeline allows you, or a designer, to visually lay out your design. In addition, it allows you to use motion tweens on the Timeline to animate properties of your images. This is usually the preferred method for images that do not need to be added or removed from the Stage at runtime.

If you have images that need to be added or removed from the display list at runtime, you can use ActionScript to display your images. In order to have your images be available to be accessed by ActionScript, you need to have them exported for ActionScript. There is a check box in the Properties dialog box of the Library image that you can select to set this.

After your image has been exported for ActionScript, you will access it just as you would any other nonvisual ActionScript class. To do so, you will need to give your image a class name. When you select the Export for ActionScript check box, a class name is set to the same name as the Library item by default. Chances are you will want to change that in order to maintain some sort of consistency with the other ActionScript class naming conventions that you have adopted.

Your image gets exported to ActionScript as a `BitmapData` class. By instantiating your class, you can add it to a `Bitmap` instance, which then can be added to the display list.

Display Images

① Right-click the image in the Library.

Note: *For details on how to import images to the Library, see the previous section, "Import Images."*

② Click Properties.

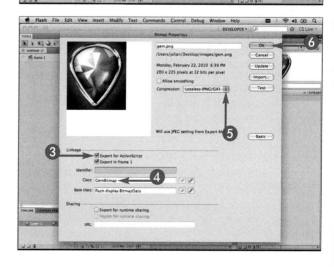

The Bitmap Properties dialog box appears.

③ Click Export for ActionScript.

④ Rename the class name, such as `GemBitmap`.

⑤ Click here and select Lossless (PNG/GIF).

⑥ Click OK.

Your image is ready to be accessed in ActionScript as a class.

116

7 Open the Actions panel.

8 Instantiate your image from the Library, such as `var bd:GemBitmap = new GemBitmap();`.

9 Add it to a `Bitmap` instance, such as `var gem:Bitmap = new Bitmap( bd );`.

10 Add the `Bitmap` instance to the Stage, such as `addChild( gem );`.

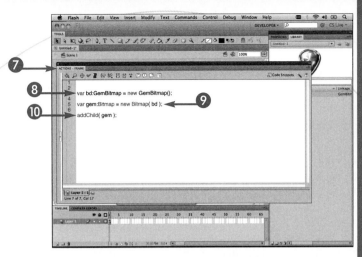

11 Press ⌘+Enter (Ctrl+Enter) to test your movie.

● Your image is now on the Stage.

Extra

Giving your image a better class name is entirely up to you. The important thing is to pick a convention and be consistent throughout your code. Adobe and the Flash community have adopted a set of naming conventions, which are consistent with those of ECMAScript. Choosing good names is very important because it makes your code clear and easy to understand. Even if you are the sole developer, chances are at some point another developer will look at your code and need to understand it.

Package names always start with a lowercase letter and continue with intercaps for subsequent words, such as `core` and `scrollClasses`. Namespaces start with a lowercase letter and use an underscore to separate words, such as `mx_internal`. Interface names start with a capital *I* and use intercaps for subsequent words, such as `IBitmapDrawable`.

Class names always start with a capital letter and use intercaps for any remaining words, such as `Accelerometer` and `CameraRoll`. Event and Error subclasses should following Adobe's naming convention and place the word at the end of the name, such as `AccelerometerEvent` and `IOError`.

Bundle Images with Your Application

Importing your images to the Library, which bundles them inside your application binary, is one way to bundle images with your application. For more details on this topic, see the section "Import Images" earlier in this chapter. The other method is to bundle the images externally from your application binary, which can be loaded and displayed at runtime with ActionScript.

This method has many benefits. First, by not including the images in the binary of the application, you reduce its size, which speeds up application load times. Secondly, the images get loaded into memory only after they have been loaded from the file system of the device. The images will also not be cached and can be freed from memory when they are no longer needed. This gives you more control over the memory consumption of your application.

This method is good for images that are not used frequently and at times when performance is not critical. Splash screens and help screens are good examples of these uses.

You can bundle files with your application from the iPhone Settings dialog box. At the bottom of the dialog box, there is a list of files that will be included with your application. By default, there are two in the list, your .swf file and your application descriptor file. You can add additional files or folders to this list.

It is a good idea to get in the habit of using relative path names for your files and folders. This will give you the ability to move your project folder to another location or computer if you are working with a team of designers and developers.

Bundle Images with Your Application

1 Click File.

2 Click iPhone OS Settings.

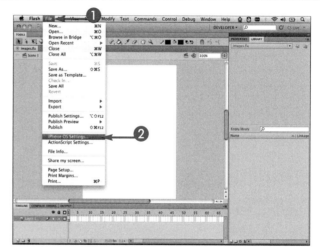

The iPhone Settings dialog box appears.

3 Click the + button to add a file.

The Open dialog box appears.

4 Navigate to and select the image to bundle.

5 Click Open.

You are returned to the iPhone Settings dialog box.

6 Click OK.

7. In a Finder window or Explorer, navigate to the folder with your compiled application.

8. Change the file extension of your compiled application from .ipa to .zip.

9. Extract the ZIP file.

10. Click here to expand the Payload folder.

11. Right-click the .app file.

12. Click Show Package Contents.

• Your image is now bundled with your application.

Extra

There is no guideline to follow in deciding which images you should import into the Library to be compiled with your file and which images you should bundle with your application. Every instance and every application are going to be different, and it will be up to you to experiment, test, and decide which method works best. The question to ask yourself is if it is acceptable for the images to take a few seconds to load and appear on the screen. The more images you load at a given time, the longer it will take for all of the images to appear on the screen. This is due to the performance requirements for loading an image into memory and displaying it on the Stage. For elements such as user interface elements and backgrounds, this is probably acceptable. However, for any kind of image sequences or game elements, this may not be acceptable. Also, any items that have to be in sync with any other element, such as sound, may be better if they are imported into the Library.

Load Images at Runtime

Loading images at runtime can be accomplished by bundling them externally from your application. For more details on how to bundle your images, see the preceding section, "Bundle Images with Your Application." Images bundled with your application will be stored in the same directory as your application.

The Loader class is used to load external assets, such as images and swf files. The Loader class extends the DisplayObjectContainer class and can be used as the parent container for the images that it loads. This allows you to add the instance of your Loader class to the display list before the asset has been completely loaded.

However, the Loader class can have only one child object, and you will not be able to add or remove any of its children as you would a normal DisplayObjectContainer instance. The Loader class exposes the loaded asset by accessing the content property. It is only accessible after the file has completely finished loading.

To make sure that the image is available, you can listen for the Event.COMPLETE event on your Loader instance. This will ensure that the file has been completely loaded and is ready to be added to a different display list.

Even though you are loading the file locally from the device's file system, loading will not happen instantaneously. Depending on your assets, loading them can be a performance-intensive process, especially if you are trying to load multiple images at the same time. This may cause your images to flicker as they are loaded. Every situation will be unique, and you will have to test and make adjustments that create the best possible solution for your needs.

Load Images at Runtime

① Create a Loader instance, such as `var loader:Loader = new Loader();`.

② Create a URLRequest instance for the image that you bundled with your application, such as `var request:URLRequest = new URLRequest( "gem.png" );`.

Note: *See the preceding section, "Bundle Images with Your Application," for more details.*

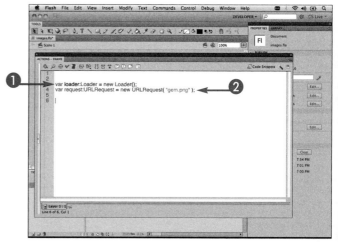

③ Listen for when the image has finished loading, such as `loader.contentLoaderInfo.addEventListener( Event.COMPLETE, onComplete );`.

④ Create an event handler for the complete event, such as `onComplete`.

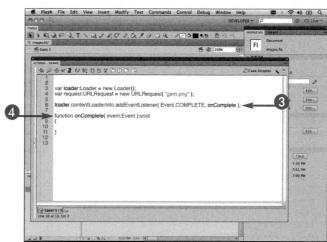

⑤ Get a reference to the loaded `Bitmap`, such as `var bitmap:Bitmap = loader.content as Bitmap;`.

⑥ Add the loaded `Bitmap` to the stage, such as `addChild( bitmap );`.

⑦ Load the image, such as `loader.load( request );`.

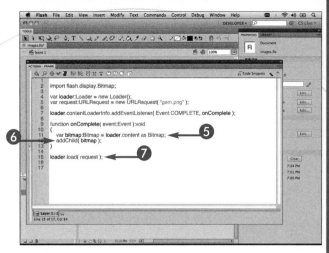

⑧ Press ⌘+Enter (Ctrl+Enter) to test your movie.

● Your image is now on the Stage.

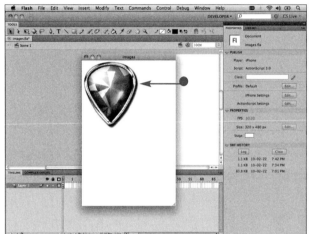

Apply It

You can also load images from the Internet just as easily as you did from the local file system of the device. The only difference will be the URL that you pass to the `Loader` instance. One thing that you will have to keep in mind, however, is the load times over a network. You will want to indicate to your user that something is loading and potentially provide them with progress updates. To do so, you can listen for the `ProgressEvent.PROGRESS` event on your `Loader` instance, as follows:

```
var loader:Loader = new Loader();
loader.contentLoaderInfo.addEventListener( ProgressEvent.PROGRESS, onProgress );
function onProgress( event:ProgressEvent ):void{
var percent:Number = event.bytesLoaded/event.bytesTotal;
}
```

Create Images Dynamically

Earlier topics in this chapter explore developing with images that you created for your application in an image-editing application, such as Photoshop. But did you know that you can also create images dynamically at runtime with ActionScript? To create an image, you can use the BitmapData class in order to take a snapshot of a particular object or region of the screen. This is almost like taking a screenshot of your application, but it gives you more control over size and location. This technique is used often in user-generated applications to save the user's creations so that they can be preserved and viewed at a later time.

The BitmapData class is the pixel representation of a Bitmap object. The BitmapData class allows you to manipulate those pixels in various ways, as well as create new ones. The draw() method allows you to take a snapshot of a source object, which can later be saved as a Bitmap. Valid source objects are any DisplayObject, such as MovieClip, Sprite, Bitmap, Video, and TextField; also, you can specify another BitmapData object as the source.

You can also alter or apply effects to the drawn representation of your source object. The draw() method takes a number of optional parameters that provide you with this functionality, such as applying a colorTransfrom and blendMode, as well as using a Matrix instance to scale, rotate, or translate the bitmap.

After your object is drawn to a BitmapData instance, you can add it to a Bitmap instance to display it on the screen, or you can encode the data to an image format such as PNG and save it to the file system.

Create Images Dynamically

1. Create a BitmapData instance, such as var bd:BitmapData = new BitmapData();.

2. Give it a width, such as 320.

3. Give it a height, such as 480.

4. To make it not transparent, enter false as the third argument.

5. Enter an ARGB value as the background color, such as 0xFFFF0000;.

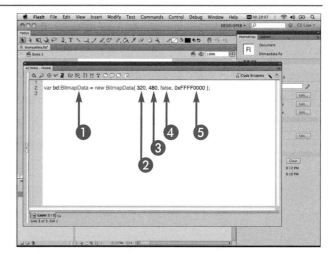

6. Create a Rectangle instance, such as var rect:Rectangle = new Rectangle(0,0,160, 480);.

7. Fill the rectangle with a color, such as bd.fillRect(rect, 0xFF00FF00);.

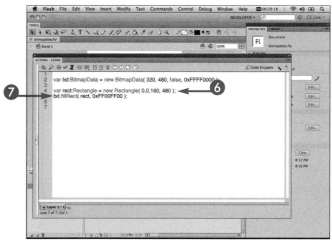

8. Create a Bitmap instance, such as var bitmap:Bitmap = new Bitmap();.

9. Add your BitmapData instance to the Bitmap instance.

10. Add the Bitmap instance to the Stage, such as addChild(bitmap);.

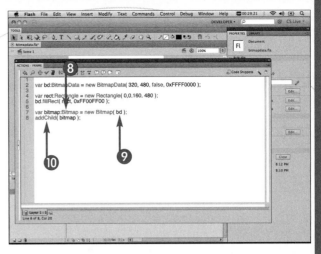

11. Press ⌘+Enter (Ctrl+Enter) to test your movie.

• Your dynamically created image is added to the Stage.

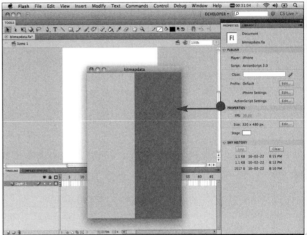

Apply It

Currently, the Flash Player does not have a method for encoding PNG images. However, there are several open source ActionScript 3.0 classes that do just that. The as3corelib, which was created by Adobe and is available at http://code.google.com/p/as3corelib/, has a PNGEncoder class that allows you to pass it a BitmapData instance. The PNGEncoder.encode method returns a ByteArray encoded in the PNG image format. This allows you to save the ByteArray to the file system of your device, as follows:

```
var bd:BitmapData = new BitmapData( 200, 200, false, 0xFFFF0000 );
var png:ByteArray = PNGEncoder.encode( bd );
```

Save Images to the Photo Library

Currently, there is no camera support for Flash CS5 iPhone applications. This means that you are not able to take a picture and save it to the camera roll as you would with a native iPhone SDK application. However, Adobe has introduced the ability to save images to the camera roll, through a new class called `CameraRoll`.

The `CameraRoll` class is a very simple class that has only one static method and one static property. The `supportsAddImage` static property can be used to check whether the platform your application is running on supports adding images to the media library. In the case of an iPhone, this should always return `true`; however, if your application is running on multiple platforms, it is a good idea to make this check.

The `addImage` static method enables you to save a `BitmapData` instance to the camera roll. There are a number of ways to create a `BitmapData` object, as shown earlier in this chapter. For example, if you are creating a painting program, in which users can paint on the screen with their fingers, you may want them to have the option to save their paintings. You could save them to the Documents directory of your application; however, this makes it a little more difficult for the users to get the image off of their device. Saving the image to the camera roll gives them more options, the easiest being to just email it to themselves from the Photo application.

After the image has been successfully saved to the camera roll, there is currently no fully supported way to load that image again. See the following section, "Load Images from the Photo Library," for more details.

Save Images to the Photo Library

1. Import an image to the Library.

   Note: See the section "Import Images" for more information.

2. Click and drag it onto the Stage.

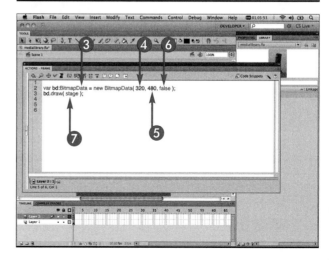

3. Create a new `BitmapData` instance, such as `var bd:BitmapData = new BitmapData();`.

4. Give it a width the same as your image, such as 320.

5. Give it a height the same as your image, such as 480.

6. Type **false** to make it nontransparent.

7. Draw the Stage into your `BitmapData` instance, such as `bd.draw( stage );`.

⑧ Check to see if the camera roll is supported.

⑨ Create a new `CameraRoll` instance, such as `var cr:CameraRoll = new CameraRoll();`.

⑩ Add the `BitmapData` to the `CameraRoll`, such as `cr.addBitmapData( bd );`.

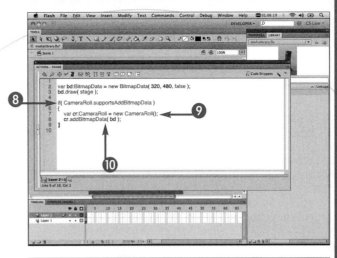

⑪ Compile your application and install it on your device.

⑫ Run your application on your device.

You can go to the Photos application to see your image.

Apply It

The `CameraRoll.addImage` static method also has a second parameter, which is a function that will be fired after the image has been successfully saved. This parameter is optional, but it is a good idea to implement it. You may want to show a progress indicator animation while it is saving. Implementing the second parameter will allow you to remove your animation and re-enable the user interface of your application.

```
var cr:CameraRoll = new CameraRoll();
cr.addImage( mybitmapdata, onComplete );
function onComplete():void {
trace( "image successfully saved" );
}
```

Load Images from the Photo Library

The preceding section, "Save Images to the Photo Library," explores how to save images to the camera roll in the Photos application. So how do you display an image from the camera roll? Currently, at the time of writing, there is no formal support from Adobe on this. The reason is that the iPhone SDK has a very specific way that a user selects an image, and creating an ActionScript API that will be consistent across multiple platforms is very challenging. However, there is an unsupported way to display images that you saved to the camera roll. I will warn you that this method may make Apple reject your application, as it is not fully supported.

When images are saved to the camera roll, they are saved to the following location on the device file system:

/private/var/mobile/Media/DCIM/100APPLE. Having this information makes it pretty easy to load the images into your application.

You do not want to rely on any specific naming convention that Apple may use, so you can simply loop over the entire directory's contents. When you do so, you can determine if the file is a valid image, a PNG or JPEG, by checking its file extension. If the file is an image, you can load it with a Loader instance, as shown in the earlier sections in this chapter.

After the images have loaded, you can size them to be 75 pixels wide and 75 pixels high and arrange them in a grid pattern. This is the same grid pattern and sizing that the iPhone SDK image picker control uses.

Load Images from the Photo Library

① Create rows and columns variables.

② Create a File instance, such as var file:File = new File("/private/var/mobile/Media/DCIM/100APPLE");.

③ Get the directory listings.

④ Create a loop to loop through the directory listings.

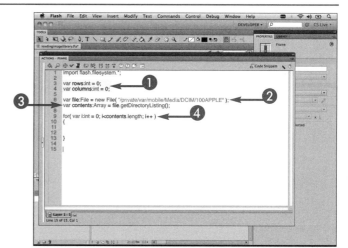

⑤ Create a File instance for the current file, such as var img:File = contents[i] as File;.

⑥ Check to see if the file extension is jpg or png.

⑦ Create a new Loader instance.

⑧ Listen for when the image has finished loading.

⑨ Load the current file.

⑩ Create an event handler for when the image has finished loading.

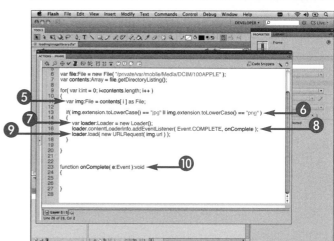

⓫ Create a MovieClip instance.

⓬ Position the MovieClip instance in a grid.

⓭ Increment the columns variable by 1.

⓮ Check to see if it is the fourth column.

⓯ Set the columns variable to 0.

⓰ Increment the rows variable by 1.

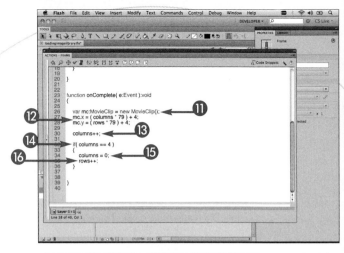

⓱ Add the loaded image to the MovieClip instance.

⓲ Set the width to 75.

⓳ Set the height to 75.

⓴ Add the MovieClip instance to the Stage.

㉑ Compile, install, and run the application on your device.

The images are loaded and displayed from the Library.

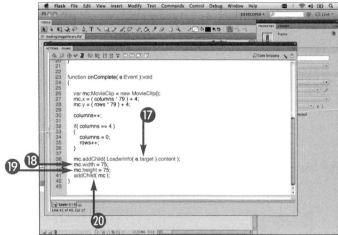

Apply It

In order to mimic the Photo application image picker, you will need to allow your users to select an image to view it full screen. To do this, you can add an event listener that listens for a touch event on the thumbnail. When it detects a touch, it would call an event handler that sets its dimensions to 320 x 480, sets its position to 0,0, and sets its index in the display list to the topmost one:

```
function onImageClick( e:MouseEvent ):void{
var img:MovieClip = e.target as MovieClip;
setChildIndex( img, numChildren - 1 );
img.x = 0;
img.y = 0;
img.width = 320;
img.height = 480;
}
```

Using iOS Default Images

Apple has provided some great images and icons in the iPhone SDK that are available only on Mac OS X. Many of these icons, such as icons for more details, deleting items, and refreshing content, are used in the built-in applications of your device.

Using these icons is a real responsibility. If you implement one of the images to do something other than what its default behavior is, you stand a good chance that you will confuse your users. There is also a good chance that Apple will reject your application for the improper use of an icon or image. The list of images is too large to describe their uses here; however, Apple has provided these in its "iPhone Human Interface Guidelines" located at the iPhone Developer Program Portal.

As mentioned earlier, these images are part of the iPhone SDK, which means that they are not readily accessible for you to include in your Flash iPhone applications. There are many Photoshop and Illustrator templates online that do a great job at replicating these images. You can, however, extract the images from the iPhone SDK if you have it installed, using the iPhoneShop 3.0 application, which can be downloaded from http://tiny.cc/iPhoneShop.

The images are stored in a file named Other.artwork, which can be found in the following directory, or one similar depending on the version: /Developer/Platforms/iPhoneSimulator.platform/Developer/SDKs/iPhoneSimulator3.0.sdk/System/Library/Frameworks/UIKit.framework/. Copy the Other.artwork file to the iPhoneShop application directory, and in a Terminal window, navigate to that directory. You will now enter the command to export the images using the Java application. After the images are exported, you can find all of them in the Other folder.

Using iOS Default Images

Note: *Before you can begin these steps, you must have first downloaded the iPhoneShop application and have the iPhone SDK installed.*

1. Navigate to the folder /Developer/Platforms/iPhoneSimulator.platform/Developer/SDKs/iPhoneSimulator3.0.sdk/System/Library/Frameworks/UIKit.framework/.

Note: *This is available only on Mac OS X, not in Windows.*

2. Copy the Other.artwork file.

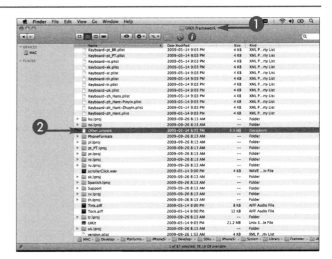

3. Navigate to the iPhoneShop folder.
4. Paste the Other.artwork file.

5 Open a Terminal window.

6 Navigate to the iPhoneShop directory.

7 Type **java –jar**.

8 Enter the jar file, such as `iPhoneShop-1.3.jar`.

9 Enter **ARTWORK** as the type.

10 Enter the artwork file, such as `Other.artwork`.

11 Enter **EXPORT**.

12 Enter the folder to which to export the images, such as `Other`.

13 Click here to close the Terminal window.

14 Open the output folder to view the files.

You can now use the images as needed.

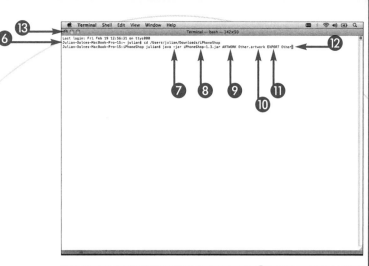

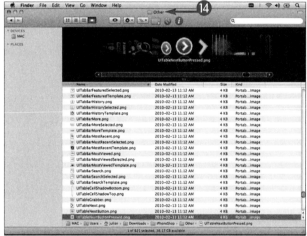

Extra

The Other.artwork file not only contains images and icons, but it also has most of the other graphics that you see for other controls in the iPhone SDK. You will be able to find images for the background to the iPhone Alert window and the background for all the different types of push buttons, scroll indicators, and progress indicator animation sequences — just to name a few.

There are additional .artwork files in the UIKit.framework folder. The Keyboard-Common.artwork file contains the graphics that make up the keyboard interface of the phone. To export the images to the KeyCommon directory, you can use the following command:

```
java -jar -iPhoneShop-1.3.jar ARTWORK Keyboard-Common.artwork EXPORT KeyCommon
```

Import Audio into Your Project

Using sounds in your application can either make or break it. Great sound design will enhance your application and provide an exceptional user experience. It can add a level of polish that can make a good application great. Bad sound design, on the other hand, can just as equally cripple an application. Repetitive sounds and unpleasant sounds can become annoying and cause users not to use your application.

There are a couple different ways to integrate sound into your applications, which will be explored throughout this chapter. The first method is to import your sounds into the Library, just as you did with images in Chapter 6, "Working with Images." Importing sounds to the Library bundles them inside your application binary. This can cause the application binary to grow in file size quickly, as sound files can be big. This is especially true for things such as background and music tracks.

When your application first loads, the entire binary for your application is loaded into memory. The bigger in file size the binary is, the longer it will take for your application to launch, and the more memory it will take to run your application. This is an important concept to understand and should be considered when importing audio files into your application. The method shown in this section is suitable for sound that you will play often, such as sound effects in a game.

When selecting your audio files for import, in the Import to Library dialog box, you choose All Sound Formats to let you know the supported types of audio files that can be imported into the Library.

Import Audio into Your Project

1. Click File.
2. Click Import.
3. Click Import to Library.

The Import to Library dialog box appears.

4. Click here and select All Sound Formats.
5. Navigate to and select a sound file, such as intro.wav.
6. Click Open.

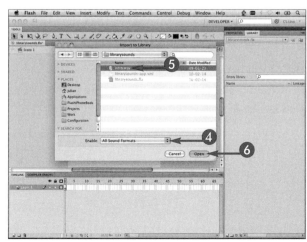

The sound file is imported into the Library.

7 Click Window.

8 Click Library.

Note: *You can also use the ⌘+L (Ctrl+L) keyboard shortcut to show and hide the Library panel.*

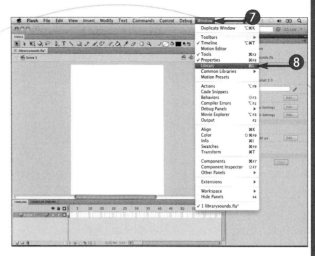

The Library panel appears.

● You can see your audio file here.

Extra

After you have imported your sound files to the Library, you can edit them in an audio-editing program, such as Adobe Soundbooth. This allows you to make edits to your sound file, such as change the volume or add filters, and have them automatically updated in Flash. To edit a sound file, right-click it in the Library and select Edit with Soundbooth. If you do not have Soundbooth installed on your machine, you can select the Edit With option instead, and it will prompt you to select your audio-editing application. Audacity is an excellent free audio-editing program that is available for both Windows and Mac OS X.

After your audio-editing program has launched, your sound file will be opened so that you can begin editing it. After your edits are complete, save the file, and your sound will be updated in Flash.

Alternatively, you can simply overwrite the file on your hard drive with a new version and select Update from the right-click menu of the sound in the Library. This allows you to update a sound file in the event that you are not the one updating the sound.

Choose an Audio Codec

Choosing the right audio codec for your sounds is extremely important. There are two different types of audio codecs: uncompressed, such as linear PCM and IMA4 (IMA ADPCM), and compressed, such as MP3 and AAC.

When you play a compressed sound, the device must first decode the sound before playing it, whereas uncompressed sounds do not require any decoding. The iPhone and the iPod can decode a compressed sound by using software decoders, as well as hardware-assisted decoders. Hardware-assisted decoders are much more efficient than software ones, and they will give you the best performance. However, all compressed sounds share a single hardware path. This means that only one sound can use the hardware-assisted decoding in order to be played back at any given time. Any additional sounds played will fall back onto the software decoder, and you will see a decrease in performance.

Given that information, if you plan on playing multiple sounds at once, such as sound effects for a game, you will want to use an uncompressed audio codec. If you are positive that you will be playing only one sound at any given time, you can use a compressed audio codec, such as MP3, without too much worry.

To use uncompressed sounds with your applications, you will need to import them into the Library of your file because Flash does not support the playback of any external uncompressed sounds. Changing the audio codec can be done from the Properties dialog box for a specific sound in the Library. There are three compressions you can choose from: ADPCM and RAW for uncompressed sounds and MP3 for compressed sounds.

Choose an Audio Codec

Set an Audio Codec for a Library Sound

1. Right-click the sound file in the Library.
2. Click Properties.

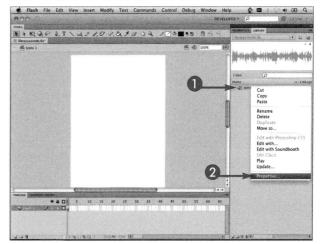

The Sound Properties dialog box appears.

3. Click here and select the codec that you want, such as ADPCM.
4. Click here to choose to convert your stereo audio to mono.
5. Click here and select a sample rate, such as 44kHz.
6. Click here and select the number of bits, such as 4 bit.
7. Click OK.

The codec that you selected is applied to the sound file.

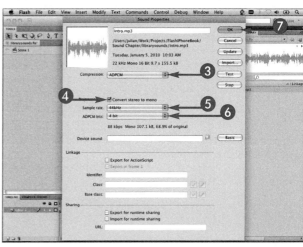

132

Set a Global Audio Codec

1. Click File.
2. Click Publish Settings.

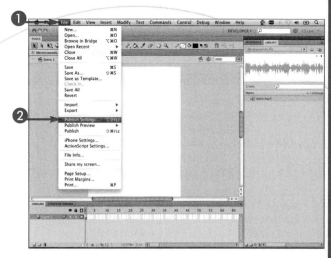

The Publish Settings dialog box appears.

3. Click the Flash tab.
4. Click the Set button for audio streams.

The Sound Settings dialog box appears.

5. Click here and select the codec that you want.
6. Click here and select a sample rate, such as 44kHz.
7. Click here and select the number of bits, such as 5 bit.
8. Click OK.

You are returned to the Publish Settings dialog box.

- You can click the Set button for audio events to set the audio compression settings.

9. Click OK.

The codec that you selected is applied to all sound files.

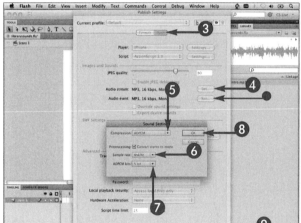

Extra

To get the most out of the audio in your application, you may want to experiment with several different audio codecs. If you have a lot of sounds in your Library, it is a pain and time consuming to change them all individually. What you may have not known is that you can select multiple sounds in the Library and change the compression of them all at once. The process for changing multiple sounds at once is the same as changing it for an individual item. This puts a big emphasis on keeping your Library organized and similar files grouped together. For example, you may want to keep all your images in one folder and all your audio files in another. This will allow you to not only find your items quickly, but also select multiple items more easily. You may also want to group items that are related to specific sections of your application, such as all your assets for the Home section in one folder, and all the assets for your games in another.

Bundle Sounds with Your Application

As well as import sounds to the Library and embed them into your application, you can bundle files externally with your application. These files will not be placed into memory when your application is launched, and you will have more control over their being released from memory when they are no longer needed.

Consider this method if you have large or long sound files. Having these not included in your application binary will help reduce the initial load time of your application. The downfall to using this method is that currently, Flash does not support playback for externally loaded uncompressed audio codecs. This leaves you with being able to bundle only MP3 files externally with your application. If you do not plan to play any of these sounds simultaneously, this method will work well. However, if you do plan for them to play simultaneously, you should look at importing the sounds to the Library. For more details on importing sounds to the Library, see the section "Import Audio into Your Project," earlier in this chapter.

You can bundle files with your application on the General tab of the iPhone Settings dialog box. When you choose the files that you want to include, it is a good practice to use relative path names. This will allow you to develop your project on a different computer or allow you to work with multiple designers and developers as part of a team.

When your application is compiled, your files will be placed in the same location as your application binary. This makes it very easy to access them with ActionScript because the base path is the exact same as your application.

Bundle Sounds with Your Application

① Click File.

② Click iPhone OS Settings.

The iPhone Settings dialog box appears.

③ Click the + button.

The Open dialog box appears.

④ Navigate to and select the file that you want to add to your bundle, such as intro.mp3.

⑤ Click Open.

134

- Your file is now added to the list of files to bundle with your application.

6 Click OK.

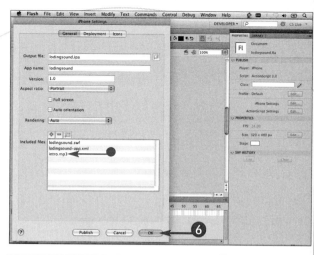

7 In a Finder window or Explorer, navigate to the folder with your compiled application.

- You can see that the audio file is bundled next to your application binary.

Extra

Bundling your sounds with your application has many benefits, such as being able to control when they are added and removed from memory. This allows you greater control over the memory footprint of your application. The disadvantage of this method is that loading sounds can be a process-intensive task, especially if you are loading multiple sounds at the same time. Even though you are going to load the sounds locally from the file system of your device, there will be a slight delay from the time you load it to the time you are able to play it. This can cause sync issues as well as some performance issues. It is recommended that you preload your sounds before they are required to be played. If you have any experience building Flash applications for the Web, you should be familiar with this approach.

As well as preloading a sound before it is needed, it is also a good idea to play it once at zero volume. This will ensure that your sound plays back smoothly when played.

Load Sounds at Runtime

Loading sounds at runtime can be accomplished by bundling them externally from your application. For more details on how to bundle your sounds, see the preceding section, "Bundle Sounds with Your Application." Sounds bundled with your application will be stored in the same directory as your application. Currently, MP3 files are the only audio format that is supported for playing back external sounds with ActionScript.

You can also load an external MP3 file from a valid URL over the network. If you plan to load audio files over the network, you should make sure that it is really apparent to the users that they are going to be doing so. Downloading audio files over 3G has the potential to be very expensive for your users, and they should have the option to opt out of that download or make sure that they are connected to a WiFi hotspot.

The Sound class enables you to load an external MP3, using the load() method. The load() method takes a URLRequest instance, which specifies the file that you want to load. After you have called the load() method and your file has begun loading, you can listen for ProgressEvents. This will allow you to monitor the progress of your file as it loads and allow you to display that to your user. Listening for ProgressEvents is really necessary only when loading sounds from the Internet, as loading sounds from the file system of the device should happen fairly quickly.

You can also listen for the Event.COMPLETE event on your sound object. This event will be fired when the sound file has been completely loaded.

Load Sounds at Runtime

1. Select a frame on the Timeline in which to place your ActionScript code.
2. Open the Actions panel.
3. Create a Sound instance variable, such as var snd:Sound = new Sound();.

Note: *This example loads a sound file from your application bundle. For more details on how to bundle sound files with your application, see the preceding section, "Bundle Sounds with Your Application."*

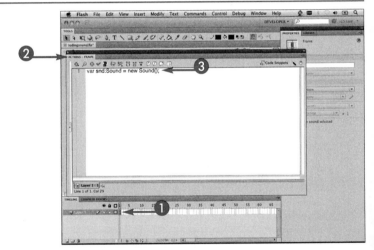

4. Listen for the ProgressEvent.PROGRESS event on the Sound instance.
5. Create an event handler function, such as onProgress.
6. Calculate the percentage of the file downloaded, such as var percent: Number = e.bytesLoaded/e.bytesTotal;.

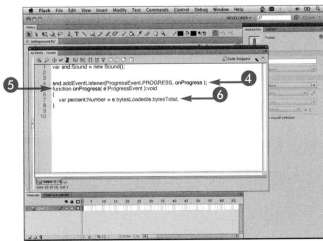

7 Listen for the Event.COMPLETE event on the Sound instance.

8 Create an event handler function, such as onSoundLoaded.

9 Create a trace statement to signal that the file has loaded.

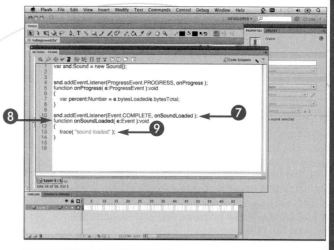

10 Create a URLRequest instance variable, such as var request: URLRequest = new URLRequest();.

11 Enter the name of the sound file that you bundled with your application, such as "intro.mp3".

12 Load the request with the Sound instance.

You can test this application on your computer before uploading it to your device. When you do so, you will see that the sound is loaded at runtime.

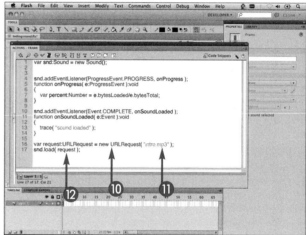

Apply It

MP3 sound files can include ID3 tags, which are metadata that is embedded in the file. *Metadata* describes the file. You can access these tags with ActionScript by listening for the Event.ID3 event on your sound object. This will fire an event when the ID3 data is available to be read when loading a sound:

```
var snd:Sound = new Sound();
snd.addEventListener( Event.ID3, onID3 );
function onID3( event:Event ):void{
var id3:ID3Info = Sound( event.target ).id3;
for (var propName:String in id3) {
trace(propName + " = " + id3[propName] );
}
```

Play Sounds

Playing sounds can be achieved through a `Sound` object instance. You can create a `Sound` object instance to load an external sound or create one from an audio file in your Library.

To make your sound files in your Library available to ActionScript, in order for you to instantiate them, you must export them to ActionScript and give them a class name. You do so using the Sound Properties dialog box. When you set the file to be exported, in the dialog box, you will see the Class text input box become active, with a default class name the same as its name in the Library. Most likely, it will be the same as the actual filename you imported and does not fit with your class naming conventions. Make sure to change that to something that reflects an ActionScript class so that you can remain consistent.

After you have created your `Sound` object instance, you can play it with the `play()` method. This method has three parameters: the start position of the sound, the number of times to loop the sound, and a `SoundTransform` object for your sound. All three of these parameters are optional and can be omitted if you want to play a sound quickly and only once.

The `play()` method also returns a `SoundChannel` instance, which allows you to control the sound. Each sound in your application is assigned to a `SoundChannel`, and each `SoundChannel` can be mixed independently from each other.

Play Sounds

1. Right-click the sound item that you want to play in the Library.
2. Click Properties.

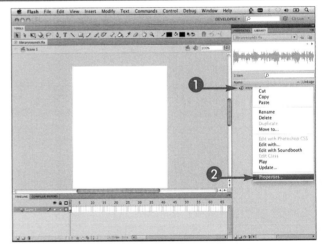

The Sound Properties dialog box appears.

3. Click Export for ActionScript.
4. Give your sound a class name, such as `IntroSound`.
5. Click OK.

Note: *For details on choosing compression settings, see the section "Choose an Audio Codec" earlier in this chapter.*

The sound is exported for ActionScript with a class name.

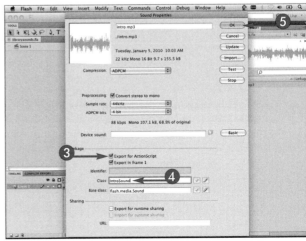

6 Click Window.

7 Click Common Libraries.

8 Click Buttons.

The Buttons Library panel appears.

9 Click a sample Play button.

10 Drag the button onto the Stage.

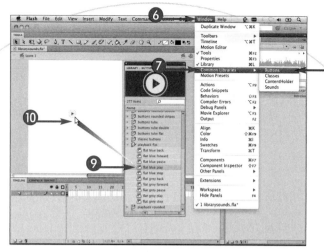

11 Create an instance of your sound from the Library, such as `var snd:IntroSound = new IntroSound();`.

12 Listen for a click event on the Play button on the Stage.

13 Create a click event handler function, such as `onPlay`.

14 Play the sound, such as `var channel:SoundChannel = snd.play();`.

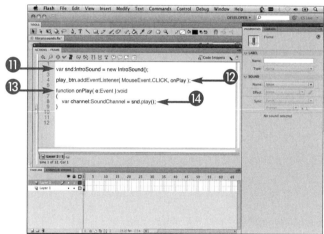

Apply It

When playing a sound, there are going to be times when you want to know when that sound has finished playing. One example for this would be to play another sound after the first one has finished. For example, you could create a bunch of small loops instead of one longer background track for the sound track of your application. This will give you lots of flexibility to have dynamic sound tracks, and they will not get repetitive and annoy the user.

You can add an event listener to the `SoundChannel` of your sound to determine when it has completed playing, such as the following:

```
var snd:Sound = new MyLibrarySound();
var channel:SoundChannel = snd.play();
channel.addEventListener( Event.SOUND_COMPLETE, onSoundComplete );
function onSoundComplete( event:Event ):void{
//sound has finished playing.
}
```

Stop Sounds

After you have started a sound using the `Sound.play()` method, you will want to stop it at some point. The first thing I am sure that you will be looking for is a `stop` method on the `Sound` class; however, this does not exist. The preceding section, "Play Sounds," discusses how the `play()` method returns a `SoundChannel` object instance. The `SoundChannel` class is what is used to control all aspects of your sound. You can think of it as a channel on a sound mixing board, with each of your sounds assigned to an input.

`SoundChannel` implements only one method, the `stop()` method. Calling the `stop()` method of a `SoundChannel` will stop the sound at the current position of the play head. It is a little confusing at first that the `play` and `stop` methods are not from the same class, but after a while, you will get used to it, and it will seem like second nature.

This method is great for stopping individual sounds, but what if you want to stop all the sounds in your application? You could put references to all your active sound channels in an `Array` and loop through them to stop them individually. As mentioned earlier, a `SoundChannel` class is like a channel on a mixing board, and the `SoundMixer` class makes that analogy a good one. The `SoundMixer` class contains static methods and properties to control sounds globally in your application. The `stopAll()` method will stop all currently playing sounds. Just like the `SoundChannel.stop()` method, this method stops the sound at the current location of the play head.

Stop Sounds

1. Click Window.
2. Click Common Libraries.
3. Click Buttons.

   The Buttons Library panel appears.

4. Click a sample Stop button.
5. Drag the button onto the Stage.

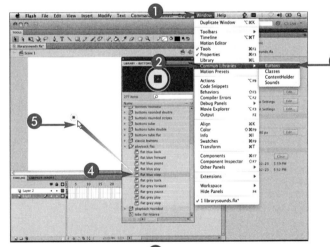

6. Select the Stop button on the Stage.
7. Click Window → Properties to open the Properties panel.
8. Give the button an instance name, such as `stop_btn`.

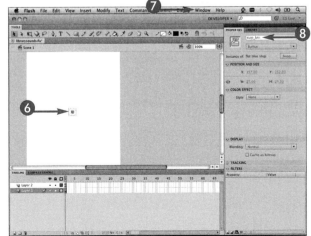

⑨ Select a frame on the Timeline in which to place your ActionScript code.

⑩ Open the Actions panel.

⑪ Create an instance variable for your sound in the Library, such as `var snd:IntroSound = new IntroSound();`.

⑫ Play the sound, such as `var channel:SoundChannel = snd.play();`.

Note: *For more details on how to play sounds, see the preceding section.*

⑬ Listen for a click event on the Stop button on the Stage.

⑭ Create a click event handler function, such as `onStop`.

⑮ Stop the sound, such as `channel.stop();`.

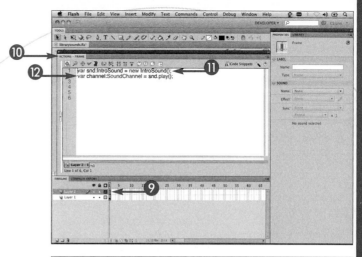

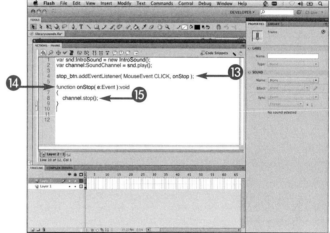

Apply It

After your sound is stopped, chances are at some point you will want to resume it from its current location. Unfortunately, there is not a `resume` method in the `SoundChannel` class for you to use, so you need to create your own. The first argument of the `Sound.play` method is the initial position in milliseconds where the sound should start to play. The `SoundChannel` class has a read-only `position` property that returns the position of the play head in milliseconds. With this, you can resume your sound from its current position, as follows:

```
var snd:Sound = new MySound();
var channel:SoundChannel = snd.play();
channel.stop();
snd.play( channel.position );
```

Set the Volume of a Sound

Setting the volume of your sound is easily done with the `SoundTransform` class. The `SoundTransform` class has a `volume` property, which determines the volume of your sound. This value can be a number between 0 and 1. Setting the `volume` property to 0 will mute the sound, and setting the property to 1 will play the sound at its loudest volume.

Setting the volume of your sounds gives you control over the entire audio mix of your application. Making sure that all of your sounds are set to the proper volume is an important part of sound design. If one sound is much louder than the others, it will stand out and cause a poor user experience. You can also increment or decrement the volume over time to fade sounds in and out.

After a `SoundTransform` instance has been created and its volume set, you can apply it to a `SoundChannel` instance through its `soundTransform` property. It is important to note that you cannot set the `volume` property of a `SoundTransform` instance that is already set on an object. For example, the following will not work:

`mychannel.soundTransform.volume = 0.5;`

Each time you would like to change the volume, you will need to reset the `soundTransform` property for the change in volume to take effect.

You can also apply a `SoundTransform` instance to other objects that have a `soundTransform` property. For example, you can set a `SoundTransform` instance to the `soundTransfrom` property of a `MovieClip` in order to control the volume of a sound that has been placed on its Timeline.

Set the Volume of a Sound

1. Create a `Sound` instance from a sound in the Library.

2. Create a `Number` instance variable, such as `var volume:Number = 0;`.

3. Create a `SoundTransform` instance variable, such as `var sndTransform: SoundTransform = new SoundTransform();`.

4. Set the volume of your `SoundTransform` instance variable.

Note: *For details on how to export a sound for ActionScript, see the section "Play Sounds."*

5. Play the sound, such as `var channel:SoundChannel = snd.play();`.

6. Set the start offset of the sound, such as `0`.

7. Set the number of times to loop the sound, such as `100`.

8. Set the initial `SoundTransform` of the sound.

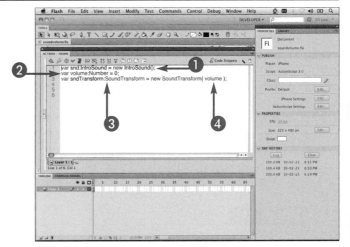

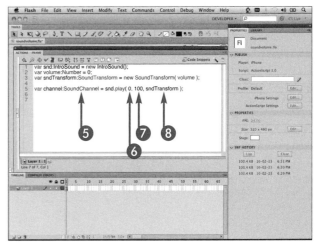

9 Add a listener to the `Event.ENTER_FRAME` event of the stage.

10 Create an event handler function, such as `fadeVolume`.

11 Increment the `volume` instance every frame.

12 Check to see if the `volume` is greater than 1.

13 Remove the `Event.ENTER_FRAME` event.

14 Constrain the volume to 1.

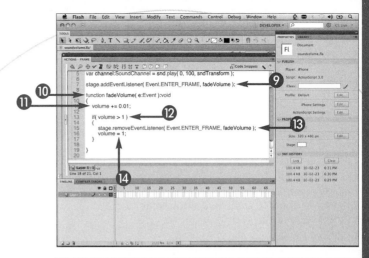

15 Create a `SoundTransform` instance variable, such as `var sndTransform:SoundTransform = new SoundTransform();`.

16 Set the volume of your `SoundTransform` instance variable.

17 Apply the `SoundTransform` instance to the `SoundChannel` variable returned from the `snd.play()` method.

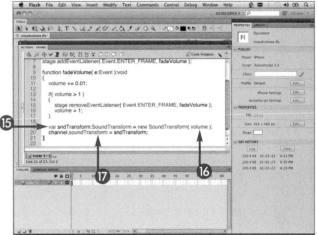

Apply It

In the section "Stop Sounds," the `SoundMixer` class was introduced. The `SoundMixer` class provides you with static methods and properties to control sounds globally in your application. Just like the `SoundChannel` class, `SoundMixer` has a `soundTransform` property that you can use to set the volume of all the sounds in your application, as in the following:

```
var sndTransform:SoundTransform = new SoundTransform();
sndTransform.volume = 0.5;
soundMixer.soundTransform = sndTransform;
```

Visualize the Sound Spectrum

The `SoundMixer` class has a method called `computeSpectrum`, which takes a snapshot of the current sound wave and places it into a `ByteArray` instance. This allows you to create some very cool audio visualizations of the sounds in your application. It is important to note that this takes into account every sound that is currently playing in your application. This means that you cannot visualize a specific sound if other sounds are playing at the same time.

The `ByteArray` that is produced from the `computeSpectrum` method has a fixed size of 512 floating-point values, in which the first 256 values represent the left channel and the second 256 represent the right channel. Each floating-point value in the `ByteArray` will range from -1.0 to 1.0.

To visualize the sound as it is playing, you will have to call the `computeSpectrum` method at a steady interval. For this, you can use the `Event.ENTER_FRAME` event, which gets fired every time the Flash Player enters a new frame. It is a good practice to listen for the sound complete event on your sound so that you can remove the event listener. If you do not, you will be doing a lot of unnecessary calculations, which can cause the performance of your application to decrease.

Visualize the Sound Spectrum

1. Create a `Sound` instance from a sound in the Library.
2. Play the sound.
3. Listen for when the sound has completed playback.
4. Create a listener for the `Event.ENTER_FRAME` event.
5. Create an event handler function for the sound complete event.
6. Remove the `Event.ENTER_FRAME` event listener.
7. Create an event handler for the enter frame event.
8. Create a `ByteArray` variable.
9. Create an `int` constant, such as `const PLOT_HEIGHT:int=200;`.
10. Create another `int` constant, such as `const CHANNEL_LENGTH: int = 256;`.
11. Create a `Number` variable, such as `var n:Number = 0;`.
12. Compute the sound spectrum, such as `SoundMixer.computeSpectrum (bytes, false, 0);`.

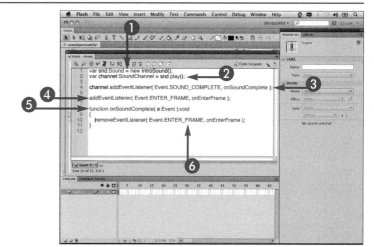

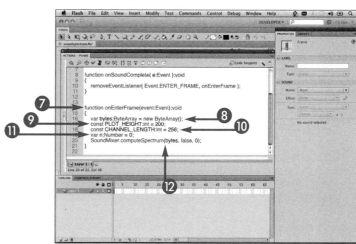

144

⑬ Create a Graphics variable, such as var g:Graphics = this.graphics;.

⑭ Clear the current graphics context.

⑮ Set the line style.

⑯ Move the drawing cursor to its start location.

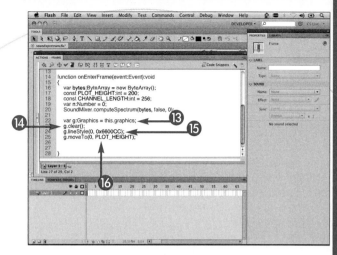

⑰ Loop through the left channel data.

⑱ Calculate the next data position of the sound.

⑲ Draw a line to the next data point.

When you publish and test your application, you will see the sound visualized.

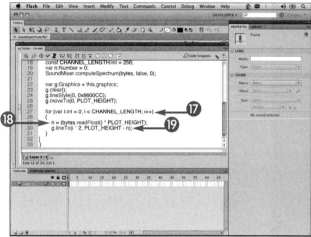

Apply It

Earlier, it was mentioned that the ByteArray created by the computeSpectrum method returned 512 values, the first 256 representing the left channel and the second 256 representing the right. In the example in this section, you only drew the representation of the left channel. You can add the following code to the end of the onEnterFrame method to draw the right channel:

```
g.lineStyle(0, 0x00FF00);
g.moveTo(CHANNEL_LENGTH * 2, PLOT_HEIGHT);
for (i = CHANNEL_LENGTH; i > 0; i--) {
n = (bytes.readFloat() * PLOT_HEIGHT);
g.lineTo(i * 2, PLOT_HEIGHT - n);
}
g.lineTo(0, PLOT_HEIGHT);
```

Explore Available Video Formats and Encode a Video File

Over the last few releases of the Flash Player, there have been many advancements in Flash Video. Currently, it supports three different codecs: Sorenson Spark, On2 VP6, and H.264. Each codec has its advantages and disadvantages; however, H.264 has become a standard for video on the Web.

Currently, Flash CS5 iPhone applications support only the Sorensen Spark and On2 VP6 video formats. These video formats can be integrated with your application a number of different ways. You can bundle them with your application as a separate file and play them with a `NetStream` object. You can progressively download a video over the network from one of your Web servers. You can use a Flash Media Server to stream the video to your device. Finally, you can embed the video on a Timeline inside your Flash file.

All of these methods produce different results, and it depends on your desired goals in order to pick what method works best for you. The following sections of this chapter discuss more about each method.

If your only option for a video codec is to use H.264, there is one way to play a video in that format. You can play the file using the native player on your device. However, this is not a recommended method, as this will cause your application to close and launch the native player. It is not ideal but is still possible using the `navigateToURL()` method, as follows:

```
var request:URLRequest = new URLRequest(
  "myvideo.m4v" );
navigateToURL( request );
```

Explore Available Video Formats and Encode a Video File

Encode a Video File

1. Open Adobe Media Encoder CS5.

Note: *You can find the encoder in the folder /Applications/Adobe Media Encoder CS5/ on Mac OS X and in the Adobe Media Encoder CS5 folder in your Program Files directory in Windows.*

2. Click Add.

   The Open dialog box appears.

3. Navigate to and select the video file to encode.

4. Click Open.

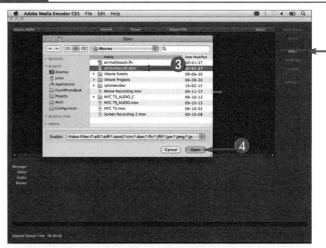

5. Click here and select a video format, such as FLV | F4V.

Note: *FLV is a Flash Video file, and F4V is a Flash renamed MP4 file. With this option, you are given the choice of several presets with different associated codecs.*

6. Click here and select an encoding preset, such as FLV – Web Medium (Flash 8 and Higher).

Note: *For more details on how to customize your encoding settings, see the following section, "Convert Videos."*

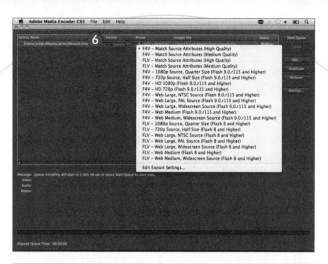

7. Select the output filename for your video.
8. Click Start Queue.
- The process of encoding your video begins.

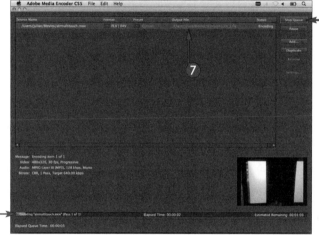

Extra

Playing video can be an expensive process on the hardware of your device. When playing video, try to have as little else going on in your application as possible. This allows the CPU to use as much processing power as it can to ensure the fastest decoding and smoothest playback possible.

Also, try to minimize the amount of ActionScript that you have executing while the video is playing. Running timers and enter frame loops as well as the Timelines will have an impact on how well your video plays back.

Try to also minimize the amount of other visual elements that are redrawn on the screen at the same time the video is playing back. This is especially true for elements that intersect or overlap the video in any way. Even if these elements are underneath the video and hidden from view, they will still be redrawn and take up precision processing power.

If you need to add controls for your video, try not to place them on top of it. Instead, place them below the video or at any location where they will not overlap. If you need to update the visual state of the controls, do so as infrequently as possible. For example, instead of updating the progress every frame, do so every second.

Convert Videos

There is a real science to converting your videos into a Flash video format. Every single video is going to compress differently at the same compression settings. Getting the best results for your videos will take some time and lots of experimenting.

The most important thing to do when converting your video is to make sure that the final video is the same dimensions as it will be when it is played back. If your application has to scale the video in order to make it fit, you will be causing unnecessary performance increases when playing it back.

There are a number of applications that will allow you to convert your video. When you installed Flash CS5, you had the option to install Adobe Media Encoder C5 as well. This application does everything you will need in order to encode your videos in any Flash video format.

After you have selected a video to encode, you can adjust the settings to best suit your needs. To encode your video in the On2 VP6 video format, be sure to select FLV on the Format tab of the Export Settings window.

On the Video tab, set the size of your video to the final dimensions that you want. If you have to resize your video, make sure to resize it in the same aspect ratio as your original. In the Frame Rate drop-down list, select the frames per second that you want, setting this to the same as your source so that the converted video will use the same FPS as your source video. The Bitrate level drop-down list has several preset values that you can use, and the Custom option will give you more control.

The Audio tab enables you to set the compression settings for the audio in your video.

Convert Videos

1. In Adobe Media Encoder CS5, click Add and choose the video to convert.

   The video is added to the queue.

2. Select your video in the list.

3. Click Settings.

Note: *For more details on how to add a video to the queue, see the preceding section, "Explore Available Video Formats and Encode a Video File."*

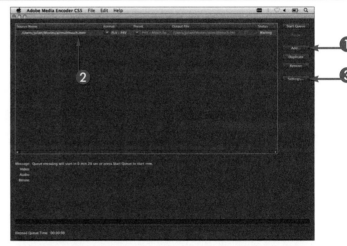

The Export Settings window appears.

4. Click here to check Export Video.
5. Click here to check Export Audio.
6. Click the Format tab.
7. Click here to select FLV.

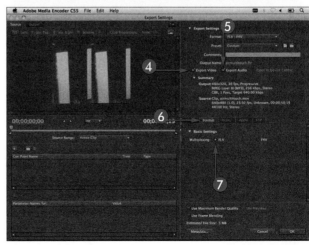

8. Click the Video tab.
9. Click Resize Video.
10. Set the width of the video, such as 480.
11. Set the height of the video, such as 320.

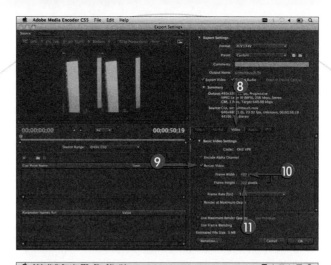

12. Click the Audio tab.
13. Select your audio to be Mono or Stereo.
14. Click here and select the bitrate of your audio, such as 128.
15. Click OK.

    You are returned to the main screen.

16. Click Start Queue.

    The process of converting your video begins.

Extra

There are a number of features in the Adobe Media Encoder application that go beyond just encoding your video. On the left side of the application, there are a number of tools that allow you to select only portions of your video to encode. You can trim the video from the start or the end to encode a subsection of your video.

You can also select the Crop tool, which is located on the Source tab, in order to crop unwanted parts of your video out. You can also make sure to crop your video to a specific aspect ratio, which helps you keep dimensions constrained.

You can also add cue points at specific times throughout your video. There are two types of cue points. *Navigation cue points* are placed on a key frame in your video. These are used for items such as bookmarks or as entry points into your video. *Event cue points* are placed at a specific time and are usually used to trigger other events in your application. To capture the cue points in ActionScript, set the `client` property on the `NetStream` object of your video.

Embed a Video

You can embed your video into your Flash file and place it on the Timeline. This allows you to control the video just as you would the Timeline of a `MovieClip`. When you import your video inside your Flash file, it will be compiled inside the binary of your application. When your application loads, it loads the entire binary into memory. This can create unnecessary memory overhead for your application, which may not be desirable. If you plan on playing your video often throughout your application, then having it always in memory may be a good idea, as loading it into memory can cause a decrease in performance.

Importing a video into your file is the same process as importing other media types, such as images and audio. However, after you select the video file that you want to import, you will be presented with several options on how you would like it imported.

The first page of the Import Video dialog box will ask you the location of your video file, and the file that you selected will be preselected. The second page of the dialog box will give you options on how you would like to embed the video in your application. There are three different symbol types to choose from: embedded video, movie clip, and graphic. Selecting Embedded Video will import your video to the Library. Selecting Movie Clip will import your video to the Library as well as create a `MovieClip` symbol with the video placed on its Timeline. Selecting Graphic will import your video to the Library and create a `Graphic` symbol with your video on its Timeline.

You also have the option to place the instance on the Stage and expand its Timeline if needed.

Embed a Video

1. In Flash, click File.
2. Click Import.
3. Click Import Video.

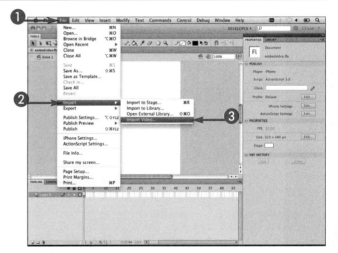

The Import Video dialog box appears.

4. Click On Your Computer.
5. Click Embed FLV in SWF and Play in Timeline.
6. Click Browse and choose the video file to import.
7. Click Continue.

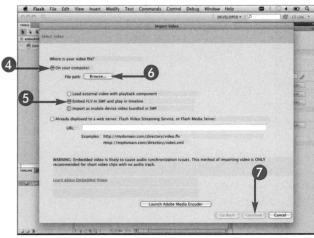

The Embedding page of the dialog box appears.

8 Click here and select a symbol type, such as Embedded Video.

9 Click Place Instance on Stage.

10 Click Expand Timeline If Needed.

11 Click Include Audio, if your video has audio.

12 Click Continue.

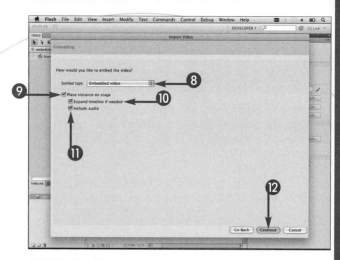

The Finish Video Import page of the dialog box appears.

13 Click Finish.

Your video is imported into your Flash file.

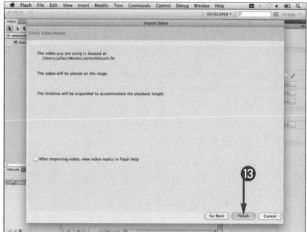

Apply It

You can also import a video into a separate Flash file and compile it as a separate .swf file. You can then load this video into your application at runtime. The one catch is that your SWF file with the video embedded in it cannot include any ActionScript. There is a restriction with loading SWF files with ActionScript. In order for this to be possible, Adobe would have to include an ActionScript interpreter with your application. This is against Apple's iPhone Developer Terms of Use and would cause your application to be rejected from the App Store. Loading your SWF file is similar to loading other assets:

```
var loader:Loader = new Loader();
var request:URLRequest = new URLRequest( "video.swf" );
loader.load( request );
```

Bundle a Video with Your Application

As well as import videos to the Library of your Flash file, you can bundle them with your application when it is compiled. This includes the file inside your application package at the same level as your application binary.

When a user loads your application on his or her device, it takes a few seconds for the application to load and start up. The time that it takes to load an application varies from application to application. When an application is loaded, iOS loads your entire application binary into memory. The bigger the file size of your application binary, the longer it will take to load, and the more memory it will use for the duration of the session.

Bundling your video with your application externally from your application binary will keep the file size of your application lower. It will also give you more control over the memory consumption of your application.

Determining which method to use when delivering and using video in your application will require some experimentation. Many games and other applications play a short video as an introduction, which is never played again. This type of video is a great example of when to bundle the video with your application.

You can add your video files to be bundled with your application on the General tab of the iPhone Settings dialog box. There are three icons above the list on this screen that allow you to add files and folders, as well as remove items from the list. Adding a folder to the list will cause its contents to be bundled with it.

Bundle a Video with Your Application

1. Click File.
2. Click iPhone OS Settings.

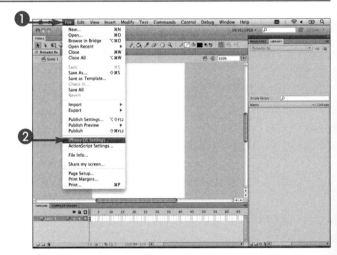

The iPhone Settings dialog box appears.

3. Click the + button.

The Open dialog box appears.

4 Navigate to and select the video file to bundle with your application, such as airmultitouch.flv.

5 Click Open.

You are returned to the iPhone Settings dialog box.

• Your file appears in the Included Files list.

6 Click OK.

When your application is compiled, your video file will now be bundled with it.

Extra

iPhone applications built with Flash CS5 have the ability to play video streams from a Flash Media Server. Streaming video files is done over the RTMP protocol. You can download and purchase Adobe Flash Media Streaming Server 3.5 from the Adobe Web site. This allows you to set up and run the server yourself. Additionally, you can install it on your development machine to use for testing for free.

If you are unable to set up and install a server on your own, there are many hosted solutions that you can use. Influxis, at www.influxis.com, is one of the premiere Flash Media Sever hosting companies available. Its staff are experts in the field and can walk you through every step in order to get your videos online.

Streaming video to the phone can use up a lot of data over the network. You will want to make sure that your users are aware that the video is streaming so that they can make sure they are connected to a WiFi hotspot. Your customers will be angry if your application uses up their data plan without their knowing it.

Load a Video

Loading a video file can be done from the file system of the device or over the Internet. This method of loading a video is sometimes referred to as *progressive download*. This is similar to how YouTube loads and plays a video.

To load a video, you will need to create an instance of a `NetConnection` object and an instance of a `NetStream` object. The `NetConnection` class is often used to connect to a Flash Remoting or Flash Media server. However, in this instance, it will not be connecting to a server, but you still need to call the `connect` method. The `NetStream` object is a channel within the `NetConnection` object, which will receive the stream data of your video.

The `Video` class is a `DisplayObject`, which is used to display the video stream data of the `NetStream` object. It can be added to the display list just like all the other `DisplayObject`s that you are familiar with, such as `MovieClip`. There is a method on the `Video` class called `attachNetStream()`, which allows it to display any video data that it receives from the `NetStream`.

To start playing your video, you use the `play()` method on the `NetStream` object. You will pass the filename of your video in as an argument into the `play()` method. Once called, your video will begin to load and start to play.

When your video starts to play, you can monitor the `bytesLoaded` and `bytesTotal` properties on the `NetStream` object. This will allow you to monitor the loading of your video and display it to the user. This is necessary only if you are loading a video from the Internet, as a video from the file system should load quickly.

Load a Video

1. Create a `NetConnection` variable, such as `var nc:NetConnection = new NetConnection();`.

2. Connect the `NetConnection` variable, such as `nc.connect( null );`.

3. Create a `NetStream` variable, such as `var ns:NetStream = new NetStream( nc );`.

4. Play a stream, such as `ns.play( "airmultitouch.flv" );`.

Note: *This will play a video that is bundled with your application.*

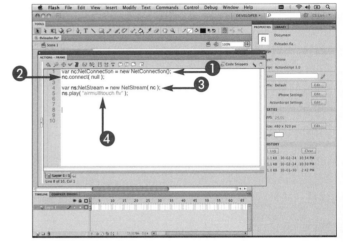

5. Create a `Video` variable, such as `var vid:Video = new Video();`.

6. Set the width of the video, such as `vid.width = 480;`.

7. Set the height of the video, such as `vid.height = 320;`.

8. Add the video to the display list, such as `addChild( vid );`.

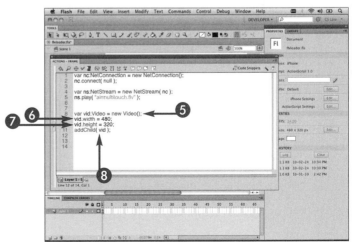

9 Set the x position of the video, such as `vid.x = 0;`.

10 Set the y position of the video, such as `vid.y = 0;`.

11 Attach the stream to the video, such as `vid.attachNetStream( ns );`.

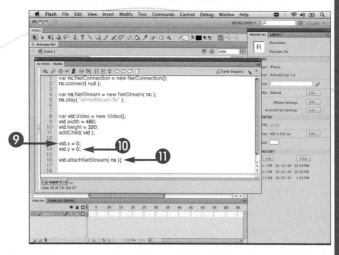

12 Press ⌘+Enter (Ctrl+Enter) to test your movie.

● The video is loaded and starts to play.

Apply It

The metadata of a video contains important information about the video. Each video encoder will embed different information into its metadata; however, most set at the very least the width, height, and duration of the video. You can retrieve the metadata of your video by creating an `onMetaData` method on the object that is set to the `client` property of the `NetStream` instance:

```
var nc:NetConnection = new NetConnection();
nc.connect( null );
var ns:NetStream = new NetStream( nc );
ns.client = this;
function onMetaData( info:Object ):void{
trace( info.width, info.height, info.duration );
}
```

If your video has metadata embedded inside of it and you do not implement this, you will receive an error similar to the following:

```
flash.net.NetStream was unable to invoke callback onMetaData.
```

Buffer a Video

When using progressive download to load a video over the Internet, you will want to implement some buffering techniques that will give your user the best playback possible. When you play a video, it does not wait for it to be loaded before it plays. The `NetStream` object fills up its buffer and then plays all the data in the buffer. As data in the buffer is played and emptied, the `NetStream` object tries to download more of the video and place it in the buffer, keeping it full. This helps deal with bandwidth fluctuation and gets your video playing as quickly as possible.

A good buffering technique to implement is called *dual threshold buffering*. This allows you to set a very low buffer time, in order to get the video playing as quickly as possible. Then when the initial buffer is full, switch to a larger buffer size to allow for a more continuous playback. This technique will allow you to deal with fluctuation bandwidth and keep your video playing as smoothly as possible.

You can listen for when the buffer is full and empty by listening to the `NetStatusEvent.NET_STATUS` event. This event is fired when there has been a change in status or error in the `NetStream` instance. There is an `info` property, which is an object, on the event that contains all the information about the new status. The `info.level` property specifies if the event is a status change or an error. The `info.code` property tells the specific event that occurred. An event with an `info.code` value of `NetStream.Buffer.Empty` means that the buffer is empty, and `NetStream.Buffer.Full` means that the buffer is full.

Buffer a Video

1. Create a `Video` variable, such as `var vid:Video = new Video();`.

2. Set the width of the video, such as `vid.width = 480;`.

3. Set the height of the video, such as `vid.height = 320;`.

4. Add the video to the display list, such as `addChild( vid );`.

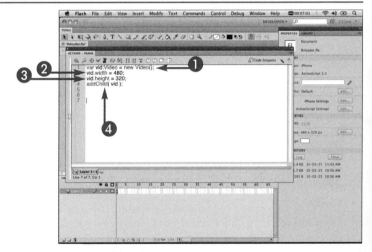

5. Create a `NetConnection` variable, such as `var nc:NetConnection = new NetConnection();`.

6. Connect the `NetConnection` variable, such as `nc.connect( null );`.

7. Create a `NetStream` variable, such as `var ns:NetStream = new NetStream( nc );`.

8. Set the `bufferTime` of the stream, such as `ns.bufferTime = 5;`.

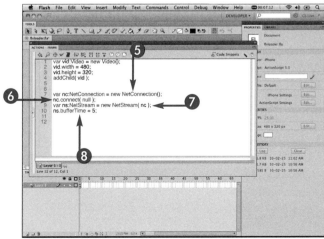

9. Set the `client` property, such as `ns.client = this;`.
10. Create an `onMetaData` function on the `client` object.
11. Listen for `NetStatusEvent.NET_STATUS` events on the stream.
12. Create an event handler function, such as `onStatus`.

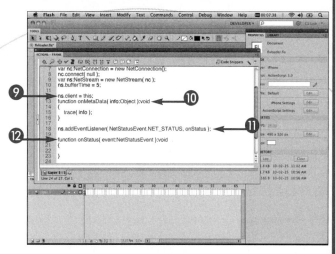

13. Create a `switch` statement for the `event.info.code` property.
14. Create a `"NetStream.Buffer.Empty"` case.
15. Create a `"NetStream.Buffer.Full"` case.
16. Play a stream, such as `ns.play( "http://www.flashiphonedevelopment.com/airmultitouch.flv" );`.
17. Attach the stream to the video, such as `vid.attachNetStream( ns );`.

When your application is compiled and tested, your video will be buffered as it plays.

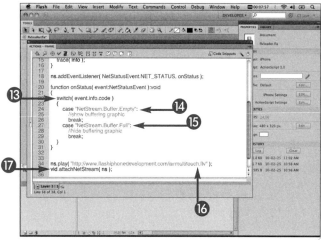

Apply It

There are many other useful messages that get sent with a `NetStatusEvent.NET_STATUS` event when fired. `NetStream.Play.StreamNotFound` is fired when you try to play a video file that does not exist. `NetStream.Play.Start` and `NetStream.Play.Stop` are fired when the stream is being started and stopped. `NetStream.Pause.Notify` and `NetStream.Unpause.Notify` are fired when the stream has been paused and then resumed. Here is an example:

```
function onNetStatus( event:NetStatusEvent ):void {
switch( event.info.code ){
case "NetStream.Play.StreamNotFound":
break;
case "NetStream.Play.Start":
break;
case "NetStream.Play.Stop":
break;
case "NetStream.Pause.Notify":
break;
case "NetStream.Unpause.Notify":
break;
}
}
```

Control a Video

After you have your video playing and buffering properly, it is time to add some controls. When creating the visual elements for your controls, make sure that they do not overlap the video in any way. Doing so will cause them to be redrawn unnecessarily and take valuable CPU processing away from the video. This section will focus on the code behind your controls; it will be up to you to implement the interface in order to affect the video.

After your video has started playing, you will want to give the user the option to pause the video or stop the video. Pausing the video can be done by calling the `NetStream.pause()` method. This stops the video at its current position, while the buffer continues to fill. Currently, there is no `stop` method in the `NetStream` class; however, you can properly implement stop functionality by pausing and then seeking to the beginning of the video.

Seeking to a specific time in the video is done with the `NetStream.seek()` method. The `seek` method seeks the video to the closest key frame at the location specified. This means that you will not always go to the exact time in seconds that you specified, but to the closest key frame before or after.

To resume a video that is currently paused, call the `NetStream.resume()` method. If the video is already playing, calling this method will have no effect.

When the user has finished watching the video and has moved onto a different section in your application, you want to make sure that you close the video stream. This will stop all data that is currently playing in the stream and make the stream available for another use, such as playing a different video.

Control a Video

1. Click Window.
2. Click Common Libraries.
3. Click Buttons.

   The Buttons Library panel appears.

4. Select buttons to use as video controls.
5. Drag them to the Stage.

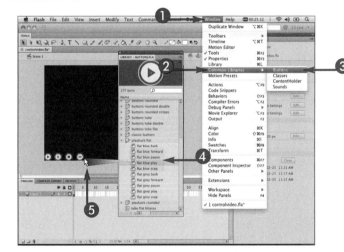

6. Select a button on the Stage.
7. Give it an instance name, such as `play_btn`.
8. Repeat steps **6** and **7** for each button.

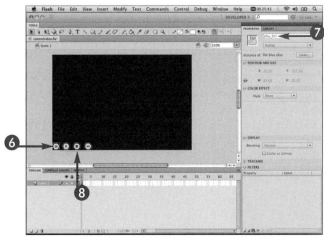

Note: *The code shown here continues the example from the section "Buffer a Video."*

⑨ In the Actions panel, add a click event listener and an event handler for the Play button.

⑩ Resume the stream, such as `ns.resume();`.

⑪ Add a click event listener and an event handler for the Pause button.

⑫ Pause the stream, such as `ns.pause();`.

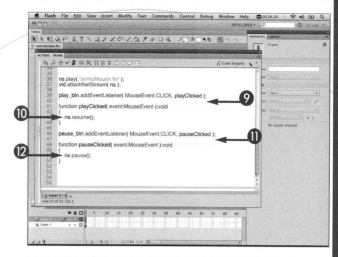

⑬ Create a click event listener and an event handler for the Stop button.

⑭ Pause the stream, such as `ns.pause();`.

⑮ Seek to the beginning of the video, such as `ns.seek( 0 );`.

⑯ Create a click event listener and an event handler for the Forward button.

⑰ Seek the video 2 seconds from its current position, such as `ns.seek( ns.time + 2.0 );`.

When your application is compiled and tested, your controls will be included with your video.

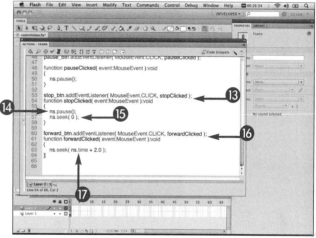

Apply It

The example in this section shows how you can pause and resume a `NetStream` instance by calling the `NetStream.pause()` and `NetStream.resume()` methods. There is also a nice convenience method implemented on the `NetStream` class called `togglePause()`. Calling this method will either pause or resume the stream, depending on what current state it is in. This makes your pause and resume logic a lot simpler, and you have to implement only one method and one button that changes its visual state when clicked:

```
function togglePause():void
{
stream.togglePause();
}
```

Chapter 8: Working with Video

Set the Volume of a Video

You set the volume of a video through the `NetStream` object. Similar to the `SoundChannel` class, the `NetStream` class has a `soundTransform` property, which can be used to affect the volume of your video.

The `soundTransform` property is set to an instance of the `SoundTransform` class. Setting the value of the `volume` property on an instance of the `SoundTransform` class will adjust the volume. This value is a number between 0 and 1. Setting the volume to 0 will mute the audio, whereas setting it to 1 will set the audio to full volume.

In order to set the volume of the `NetStream`, you must reset its `soundTransform` property every time you want to adjust the volume. This means that accessing the volume property of a previously set `soundTransform` instance will not adjust the volume of the sound that is currently playing. For example, the following code would not affect the volume of the sound track currently being played in a video:

`stream.soundTransform.volume = 0.5;`

This is true for all objects that have `soundTransform` properties, such as `MovieClip` and `SoundChannel`.

Keep in mind that some users will be using headphones when using your application. This makes it extremely important to set the initial volume of your video so that it mixes well with all the other sounds in your application. The more control that you can give the users with regards to the volume of the audio in your application, the better the users' experience will be. The last thing that you want is the users to rip their headphones off their heads because the sound is too loud or mute the sound altogether because it is poorly mixed.

Set the Volume of a Video

① Create a Volume On button symbol.

② Create a Volume Off button symbol.

Note: For more details on how to create button symbols, see Chapter 2, "Getting Started with Flash CS5."

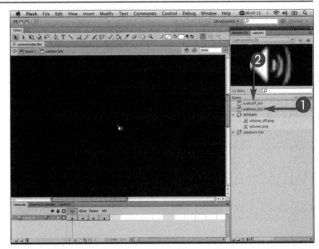

③ Place both buttons on the Stage at the same location.

④ Give them instance names, such as `on_btn` and `off_btn`.

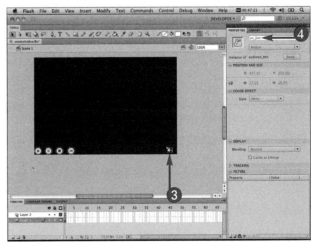

Note: *The code shown here continues the example from the section "Control a Video."*

5. Hide the On button.
6. Create a click event listener and an event handler for the On button.
7. Set the volume of the stream to 1, such as `ns.soundTransform = new SoundTransform( 1 );`.
8. Hide the On button.
9. Show the Off button.

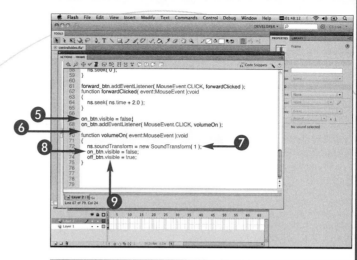

10. Create a click event listener and an event handler for the Off button.
11. Set the volume of the stream to 0, such as `ns.soundTransform = new SoundTransform( 0 );`.
12. Show the On button.
13. Hide the Off button.

When your application is compiled and tested, buttons to turn the volume on and off will be included with your video.

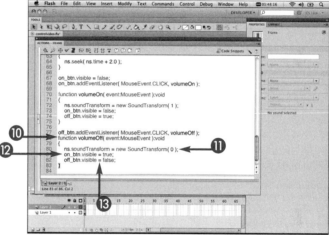

Apply It

Because users will often be using headphones when using your application, panning the audio can have a great impact on the user experience. A great example of this is the Mail application for Mac OS X. The default sound for when you successfully send an email is a swooshing sound that pans from left to right. It is not a feature that would cause you to use that email application over another, but you could argue that it definitely adds to the experience.

Setting the pan of your sounds can be done through the `SoundTransform.pan` method. This method functions in the same manner as the `volume` property. However, the pan property values range from -1 to 1. A value of -1 pans a sound fully to the left speaker, a value of 1 pans a sound fully to the right speaker, and a value of 0 sets the pan of the sounds to be in the center.

Here is an example of setting the audio of a video to be panned fully to the left speaker:

```
var transform:SoundTransform = new SoundTransform();
transform.pan = -1;
ns.soundTransform = transform;
```

Determine Available Fonts on Your Device

iOS comes with a library of fonts, also known as *device fonts,* just as your operating system for your computer does. Currently, there is not a way to determine which fonts are available with ActionScript, but there is a way to do this with Objective-C and the iPhone SDK.

With this functionality, there are several applications in the iTunes App Store that display every font that is installed on your device. One of the best applications is the Fonts application by AppEngines, http://appengines.com/fonts.html. It will display all the font families installed on the device, as well as allow you to see all the available characters for a specific font. It is a free application, and it will come in very handy as you develop applications for the iPhone and iPod.

If you are unsure of which font to use, Helvetica is usually a safe bet. When choosing a font for any input text, make sure to use one of the fonts on the device.

If you want to use a serif type font, which has semistructural details on the ends of the strokes that make up the character, you can use Times New Roman, Georgia, or _serif. If you want to use a sans-serif type font, which does not have these details on the ends of the strokes, you can use Helvetica, Arial, Verdana, Trebuchet, Tahoma, or _sans. In most fonts, each character is a different width, which can make it hard to predict how long your text will be. The contrast to this is a fixed-width font, in which each character occupies the same amount of horizontal space. If you are looking to use a fixed-width font, you can select a font from Courier New, Courier, and _typewriter.

Determine Available Fonts on Your Device

1. Go to the App Store and download and install the Fonts application on your device.

2. Scroll through the list to see all the available fonts on the device.

3. Select a font to see the whole family, such as Arial.

The whole family of the font appears.

4. Select a font to preview it on the device, such as ArialMT.

Sample text appears in the font that you selected.

5 Drag the slider to preview the font at different sizes.

6 Select the Edit tab.

7 Press inside the text area to give it focus.

8 Type a new phrase to preview the font.

Apply It

If you are developing on a computer that has Mac OS X and the iPhone SDK installed, it is really easy to write a small application to show the available fonts. The following Objective-C code will output every font family that is installed on your device to the console window. Start a new iPhone application from any of the templates in Xcode and place the code in the following method of your application:

```
- (void)applicationDidFinishLaunching:(UIApplication *)application
```

Running your application in the iPhone Simulator will return the same fonts that are installed on your device:

```
NSArray *families = [UIFont familyNames];
for( int i= 0; i<[families count]; i++ )
{
    NSLog( @"%@", [families objectAtIndex:i] );
}
```

Embed Fonts in Your Application

There will be instances in your application when you want to use a font that is not included with your device. In order for your text to appear properly, you will need to embed the font in your application. You need to specifically embed fonts only when they are used in dynamic text fields, as fonts in static text fields are automatically included with your application.

The process for embedding fonts in Flash CS5 has greatly improved. Previously to Flash CS5, fonts were embedded on a dynamic or an input text field. If you accidentally embedded extra characters that you did not need, you would have to track down the text field in your file and adjust it. Flash CS5 now has a Font Embedding dialog box that allows you to manage all of your embedded fonts from a single location.

From this dialog box, you can select fonts to embed in your application, as well as its character set. It is important to embed only the characters that are required for your application because the more characters you embed, the bigger the file size of your application will be.

You can also export your fonts so that you can access them from ActionScript. This allows you to set the font of a `TextField` instance dynamically.

When choosing a font to use in your application, it is important to understand the font license agreement of that font. The Apple iPhone SDK terms-of-use agreement states that you must own or have sufficient rights to use all assets that are included with your application. If you have downloaded a font to use from the Internet, make sure to check that you have the proper permissions before using it.

Embed Fonts in Your Application

Embed a Font in Your Application

1. In Flash, click the Text tool.
2. Click here and select Classic Text.
3. Click here and select Dynamic Text.
4. Drag a text field on the Stage.

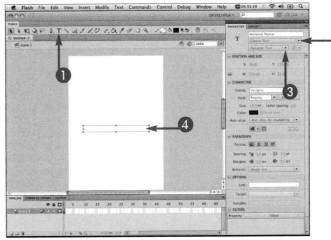

5. Click here and select a font, such as Helvetica.
6. Click Embed.

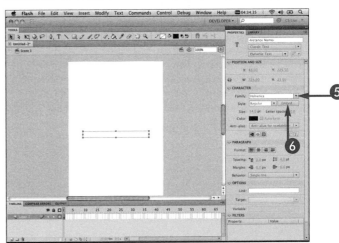

164

The Font Embedding dialog box appears.

7 Click Uppercase.

8 Click Lowercase.

9 Click Numerals.

10 Click Punctuation.

The font will now be embedded with your application when it is compiled.

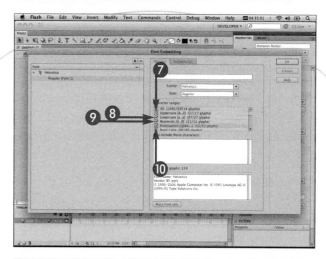

Export for ActionScript

11 Click the ActionScript tab.

12 Click Export for ActionScript.

13 Give the font a class name, such as Helvetica.

14 Click OK.

The font can now be accessed and used with ActionScript.

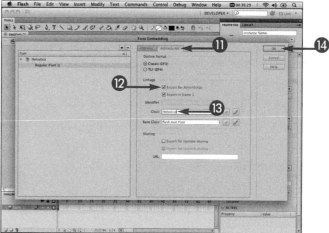

Extra

The new Font Embedding dialog box is a big improvement in managing the fonts and characters that are embedded in your application. There is, however, another way to a summary of every single character for every font that is embedded in your final application. In the Publish Settings dialog box on the Flash tab, there is Generate Size Report check box. Selecting this will generate a report of all the assets in your file and how much of the final file size of your application they account for. When your application is compiled, the report can be viewed in the Output panel. A text file is also created in the same directory as your SWF file, which allows you to compare reports as you develop.

In the Fonts section, you will be able to see each character that you have embedded for a specific font. This allows you to make sure that you are embedding only the characters that you need, which gives you greater control over the final file size of your application.

The size report also shows you the file sizes of all your images, sounds, and embedded videos.

Create an Input TextField

Adobe has done a great job of creating the same experience for the users when they are inputting text in a Flash CS5 iPhone application compared to that of a native iPhone application. Any input text field in your application will have all the same features by default, such as predictive text and the X button to clear the text field.

When selecting a font for your input `TextField`, make sure to use one of the device fonts. This will ensure that your fonts are rendered properly when inputting text. For more details on device fonts, see the section "Determine Available Fonts on Your Device" earlier in this chapter.

If you want the predictive text to appear, make sure that your text field is high enough for it to be visible.

When a user touches an input field in order to start typing, the keyboard will automatically appear. If your content is at the bottom of the screen, your content will automatically be pushed up in order to show it above the keyboard. When your user is finished inputting text, the keyboard will disappear, and your content will be readjusted to its original position. This is convenient because it makes sure that your content is always visible to your user, without your having to manage it.

Currently, you cannot control the type of keyboard that appears for a specific `TextField`. You may have noticed with some iPhone applications that the SDK has several different keyboard types, such as a number pad, a telephone pad, an email keyboard, and a URL keyboard, just to name a few. Currently, Flash CS5 iPhone applications by default use the default alphabet keyboard. You also cannot specify the auto correction type or auto capitalization type for your `TextField`.

Create an Input TextField

1. Click the Text tool.
2. Click here and select Classic Text.
3. Click here and select Static Text.
4. Create a label on the Stage.
5. Type text for your label, such as username:.

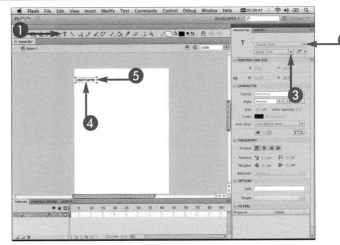

6. Click the Text tool.
7. Click here and select Classic Text.
8. Click here and select Input Text.
9. Create an input field on the Stage.
10. Give it an instance name, such as `username_txt`.

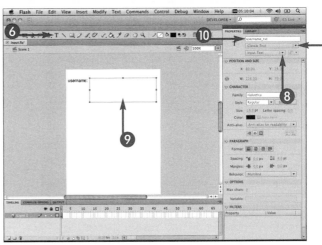

⓫ Click here and select a device font, such as Helvetica.

⓬ Click here and select Use Device Fonts.

⓭ Click the background button to create a border around your text field.

The text field is set to receive user input.

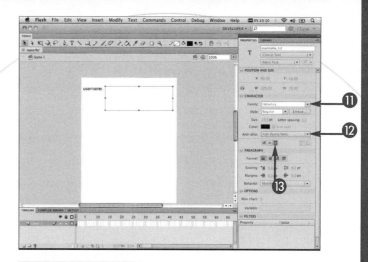

⓮ Install the application on your device.

⓯ Press the text area to give it focus and bring up the keyboard.

⓰ Type a word to see the predictive text appear.

Extra

There are a few methods that you can implement in order to hide the keyboard. The keyboard type that is used for Flash CS5 iPhone applications has a Return button as its return key. You may have seen other applications use different types of buttons in its place; however, you do not have control over this property in Flash. When the user taps the Return button, the keyboard automatically closes, and your content returns to its original position.

The keyboard will also be hidden when the user touches the screen anywhere outside the bounds of an input TextField. This is not very intuitive for users and is not standard practice for native iPhone applications. However, you can take advantage of this by creating a visual element, such as a cancel button, that when pressed by the user will hide the keyboard. The interesting thing about this method is that you do not need to create any listeners for that button in ActionScript in order to close the keyboard. When the user touches the screen over the button, the keyboard will automatically hide.

Create a Password TextField

One of the most common items that a user will input into your application is a login credential. This usually includes a password field, in which the user's input is often masked for security reasons. This prevents somebody from getting a hold of your device and being able to see your password if it is stored somewhere on your device.

When the user types into a password field, the last character that was pressed will show unmasked for a certain period of time or until he or she presses another character. This allows the user to make sure that he or she hit the correct character on the on-screen keyboard. Password text fields do not use the auto correction feature of your device.

Setting the `displayAsPassword` property on the `TextField` class enables you to hide the input characters by using asterisks or circles instead of showing the characters. When a `TextField` is set to be a password-enabled `TextField`, the copy commands will be disabled. This is a security measure to stop people from retrieving passwords from a device that is not theirs.

When allowing your users to log in in your application, it is a good practice to allow them to have their credentials stored on the device. This would allow the application to automatically log the users in, saving them from having to type their username and password every time that they start the application. There are several ways to store this information on the device, which will be explored in more detail in Chapter 10, "Saving State." However, the best method is to create a Settings bundle so that the users can change their login information from the Settings application on the device and not have to launch the application. For more details on this method, see Chapter 16, "Creating Application Settings."

Create a Password TextField

Note: You can combine the steps in the preceding section, "Create an Input TextField," with this one to create a login form.

1. Click the Text tool.
2. Click here and select Classic Text.
3. Click here and select Static Text.
4. Create a label on the Stage.
5. Type text for your label, such as password:.

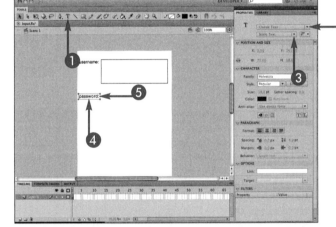

6. Click the Text tool.
7. Click here and select Classic Text.
8. Click here and select Input Text.
9. Create an input field on the Stage.
10. Give it an instance name, such as password_txt.

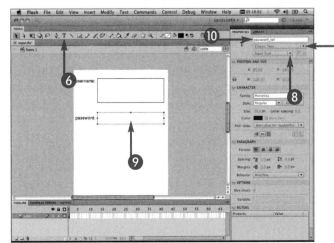

168

11 Click here and select a device font, such as Helvetica.

12 Click here and select Use Device Fonts.

13 Click the background button to create a border around your text field.

The text field is set to receive user input.

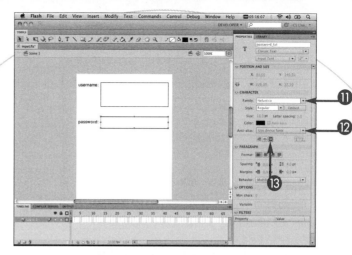

14 Click the New Layer button.

A new layer is added to the Timeline.

15 Select a frame to which to add ActionScript.

16 Open the Actions panel.

17 Set the text field to display as a password, such as `password_txt.displayAsPassword = true;`.

The password characters will now be hidden as they are typed.

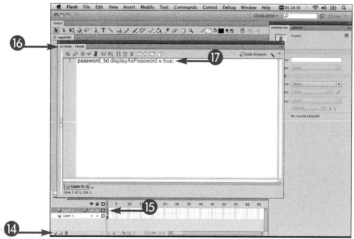

Apply It

If you are creating a form that has several input fields, you may want to give the user the ability to go to the next field without pressing it to give it focus. This is especially true when the keyboard is visible and hiding the next input `TextField`. Most iPhone applications that give users this ability supply a Next button directly above the keyboard. Clicking the Next button will force the focus to the next `TextField` in the form. To change the focus from one `TextField` to another, set the `focus` property of the `stage`:

```
function onNextButtonClicked( event:TouchEvent ):void
{
stage.focus = nextTextField_txt;
}
```

Without a Next button, the user will have to click outside of the text field area to hide the keyboard to select the next input text field. This can be very annoying if there are many fields and will cause the user to take longer to completely fill out your form.

Using TLF TextFields

Adobe has created a new Text Layout Framework (TLF), which is an extensible ActionScript library built on top of the new text rendering engine in Flash Player 10. The text engine solves many of the problems that Flash has had with text.

The new framework supports bidirectional text, vertical text, and over 30 writing systems, including Arabic, Hebrew, Chinese, Japanese, Korean, Thai, Lao, and the major writing systems of India. TLF also gives you greater control over kerning, ligature, case, digit case, and digit width.

You are also able to create additional text containers or columns and have your text flow from one to the other automatically and flow around inline images. TLF also supports proper copying and pasting and undoing when editing text. This is not the same as the method used by iPhone applications, as this is available on all text field types.

Previously, there was a major workflow issue with fonts between the Mac OS X and Windows development environments. Sometimes your text fields would shift position in your FLA file when opened on Mac OS X. The new text engine has solved these issues to allow for a more consistent workflow between platforms.

Because the Text Layout Framework is written in ActionScript, you could see some performance decreases when using some of its advanced features. When using the new text framework, your file links to the textLayout.swc file, which contains all the ActionScript code for the framework. Because Flash CS5 iPhone applications cannot include any interpretive code, you must select the framework to be merged with your code when your application is compiled.

Using TLF TextFields

1. Click the Text tool.
2. Click here and select TLF Text.
3. Click here and select a type, such as Read Only.
4. Draw a text field on the Stage.
5. Type some text, such as Hello World.

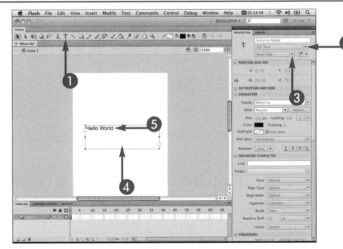

6. Click here and select Use Device Fonts.
7. Select an alignment.

   The TLF text field is created.

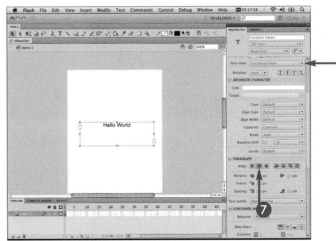

8 Click File.

9 Click ActionScript Settings.

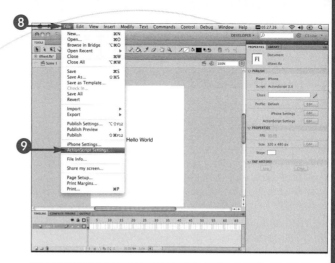

A dialog box appears.

10 Click the Library Path tab.

11 Click textLayout.swc in the list.

12 Click here and select Merged into Code.

13 Click OK.

The Text Layout Framework will be merged with your code.

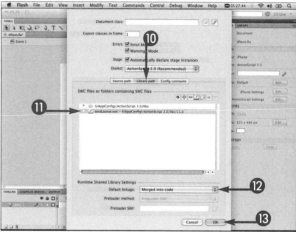

Extra

The new Text Layout Framework has provided developers and designers with a rich set of text controls in order to create the next generation text-based applications. This could not have come at a better time because the mobile reader market is about to explode, if it has not already. Devices such as the Amazon Kindle, Sony Reader, and Barnes & Noble nook already have a foothold in the mobile reader marketplace. However, new devices such as the Apple iPad are destined to take the industry to the next level. There are also a number of tablet computers set to release in 2010 that will run Adobe AIR applications and Flash Player 10.1.

Content makers have already begun developing digital versions of their content. The *New York Times* has released an Adobe AIR application that allows users to read the latest articles online, in a format that is similar to what they would expect from the printed version. This experience would not be possible without the new columns and text flow containers in the Text Layout Framework. With the new mobile devices running AIR, the *New York Times* will be able to capitalize on another customer touch point and revenue stream.

Create a Scrollable TextField

The iPhone has very limited screen real estate, and there will probably be times when you have a large piece of text that will need to be scrolled. The Flash Player for the desktop automatically scrolls the text field when you scroll the mouse wheel, when the mouse is over the top of the text field. Because there is no mouse wheel support on the iPhone, you will need to implement other methods.

One method would be to add all the text to the text field and change its height based on the amount of text. If the text field was a display container, you could move the container on the y axis as the user's finger moves up and down.

Another way would be to change the `scrollV` property of the text field based on the current direction that the user is scrolling. The `TextField.getLineIndexAtPoint()` method allows you to find the current line of text underneath a given x and y location. Using this method to track which line of text the user's finger is currently over allows you to determine which direction he or she wants to scroll the text. Comparing the current line that the finger is under with the last line from a previous frame will give you the difference in lines of text. Adding the difference to the current `scrollV` value will move the text up or down the same amount of lines as the difference.

This method allows you to have large amounts of text in a small text field and does not require more display objects or complicated caching methods.

Create a Scrollable TextField

1. Create a multiline text field on the Stage.

   Note: See the section "Create an Input TextField" for more information.

2. Give it an instance name, such as `text_txt`.

3. Open the Actions panel.

4. Create an `int` variable, such as `var startLine:int;`.

5. Create another `int` variable, such as `var lastLine:int;`.

6. Set the text field to a large string.

7. Add a mouse down listener to the text field, such as `text_txt.addEventListener( MouseEvent.MOUSE_DOWN, onPress );`.

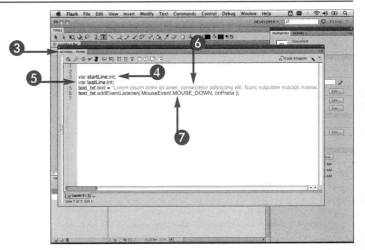

8. Create an event handler method, such as `onPress`.

9. Set the start line, such as `startLine = text_txt.getLineIndexAtPoint( e.localX, e.localY);`.

10. Set the last line to the start line, such as `lastLine = startLine;`.

11. Listen for the mouse up event on the text field, such as `text_txt.addEventListener( MouseEvent.MOUSE_UP, stopScrolling );`.

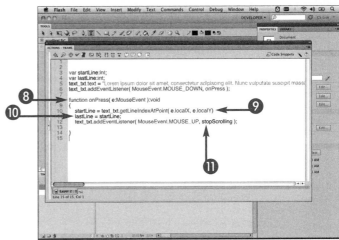

⑫ Add an enter frame listener, such as `addEventListener( Event.ENTER_FRAME, onFrame );`.

⑬ Create an event handler, such as `stopScrolling`.

⑭ Remove the enter frame event listener.

⑮ Create an event handler, such as `onFrame`.

⑯ Get the text line under the finger, such as `var currentline = text_txt.getLineIndexAtPoint( text_txt.mouseX, text_txt.mouseY );`.

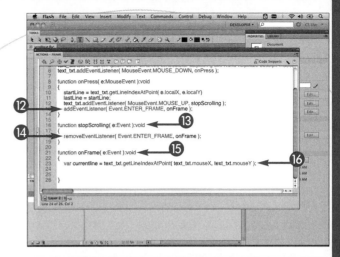

⑰ Get the line difference between the current frame and the last frame.

⑱ Scroll the text, such as `text_txt.scrollV += diff;`.

⑲ Set the last line, such as `lastLine = currentline + diff;`.

When your application is compiled, the text field will be scrollable.

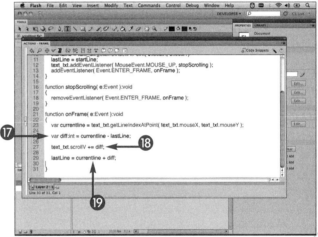

Apply It

There are a few more cases in which you do not want to scroll the text. The `getLineIndexAtPoint()` method will return -1 if the point is not over the text field. If this is the case, you want to remove the event listener for the enter frame event. Also, you do not want to scroll the text if you are at the top or at the bottom of text. Adding the following code to the `onFrame` method after the `diff` variable will solve for these cases:

```
if( currentline == -1 ){
removeEventListener( Event.ENTER_FRAME, onFrame );
}
if( diff > 0 ){
if(text_txt.bottomScrollV==text_txt.maxScrollV+text_txt.numLines) return;
}
else if( diff < 0 ){
if( text_txt.scrollV == 0 ) return;
}
```

Create a Local SharedObject

Local shared objects, often referred to as *Flash cookies*, are data files that are created by Flash applications and stored locally on your machine. Flash Web applications and sites use local shared objects to store user information so that it can be retrieved on the user's return. Unlike normal browser cookies, local shared object files are encoded in the ActionScript Message Format (AMF). This allows you to save complex objects such as `ByteArray` and does not limit you to simple text strings.

The ActionScript Message Format is a proprietary format created by Adobe that is used to serialize ActionScript objects so that they can be passed back and forth between other clients or servers. Flash applications use AMF when communicating with a Flash Remoting Server and Flash Media Server and to another Flash application through a `LocalConnection` instance.

Creating a local shared object in a Flash iPhone application creates a file in a series of folders under your application's Library/Application Support folder. The name of the file is determined by the unique name that you entered in the `SharedObject.getLocal()` method. If the file already exists, the data from the file is returned to your `SharedObject` instance; if not, an empty file is created for you.

When you are creating a name for your local shared object, most characters are supported, with the exception of spaces and these specials: ~, %, &, \, ;, :, ", ', ,, <, >, ?, and #.

If, for whatever the reason, a local shared object cannot be created by the `getLocal()` method, an `Error` will be thrown. It is good practice to surround the instantiation of your local shared object with a `try ... catch` statement in order to properly handle the error if it is thrown.

Create a Local SharedObject

Create a Local SharedObject

1. Select the frame on the Timeline to which to add ActionScript code.

2. Open the Actions panel.

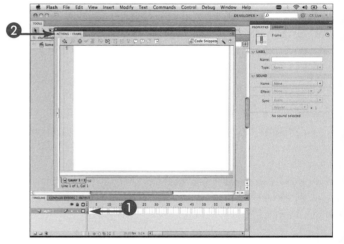

3. Create a `SharedObject` instance, such as `var so:SharedObject = SharedObject.getLocal();`.

4. Type a name for your `SharedObject`, such as `"scores"`.

   The local `SharedObject` is created.

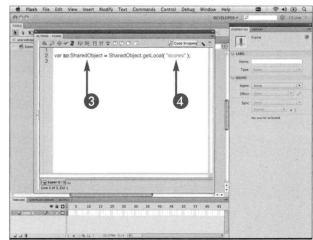

Download Your SharedObject

1. Install the application on your device and run it.
2. Open Xcode from the /Developer/Applications folder on Mac OS X.
3. Click Window.
4. Click Organizer.

The Organizer window appears.

5. Click your device's name.
6. Find the application in the Applications list.
7. Click the down arrow to download the application data.
8. Expand the folders to find your .sol file.

You can open the .sol file in a text editor to see the encoded data.

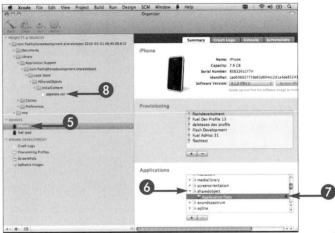

Extra

When customers download an update of your application, they will expect that any data that was saved from the previous version will still be available to them. When an update to an application is installed on the device, it is installed in a new application directory. It then moves all the user data from the old installation directory to the new one. During an update installation, the following directories are guaranteed to be copied over to the new installation application directory: <Application_Home>/Documents and <Application_Home>/Library/Preferences.

Other folders are often copied over in addition to these directories. However, because a local shared object is not stored in one of these folders, there is a chance that it may not be present after an update. It has been my experience that local shared objects are always copied during the installation of an update, and you should feel confident that they will be in your application as well.

Write to a SharedObject

After you have created your local shared object, you can write data to it. A `SharedObject` instance has a `data` property, which stores all the data in your local shared object. The `data` property is a simple `Object` and contains a collection of attributes and properties that you would like to save. These properties can be of any ActionScript type, such as `ByteArray`, `XML`, `Array`, `Number`, and `Boolean`.

Each piece of data that you want to save in a local shared object must be set as a property of the `data` property object. Directly assigning the `data` property to another object, such as `myso.data = myobject;`, will be ignored, and your data will not be saved.

After you have placed all the data that you want to save in your `SharedObject` instance `data` property, you can call the `flush()` method in order to have it immediately saved to the file system of your device. If you choose not to use this method, the data will be written to the file system when the shared object session ends. This can occur if the `SharedObject` instance is garbage-collected because there are no longer any references to it or when your application closes. Be careful when relying on this method, as there are instances, such as your application crashing, in which your data may not be saved.

Writing data to the local file system of the device can be a performance-intensive task if done in short intervals. If you can avoid calling the `flush()` method every time that you store new data in the `data` property, you will see some performance gains.

Write to a SharedObject

1. Select the frame on the Timeline to which to add ActionScript code.
2. Open the Actions panel.

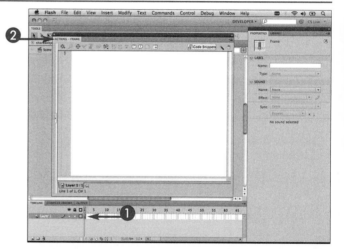

3. Create a `SharedObject` instance, such as `var so:SharedObject = SharedObject.getLocal();`.
4. Type a name for your `SharedObject`, such as `"scores"`.

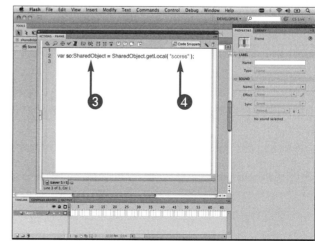

5 Store a String variable in the SharedObject, such as so.data.name = "julian";.

6 Store an int variable in the SharedObject, such as so.data.score = 1580;.

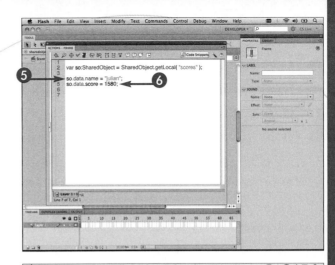

7 Create a String variable, such as var status:String;.

8 Flush the SharedObject instance to save the data, such as =so.flush();.

The data is written to the SharedObject on the file system.

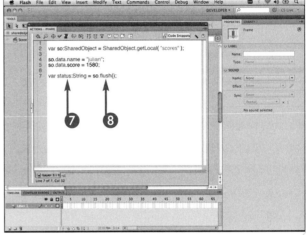

Extra

In addition to writing to a local shared object, you can delete any data that you have stored. If you want to delete a single property, you can use the delete keyword. Setting the property to null or undefined will not delete the value from the local shared object.

```
delete so.data.myattribute;
```

If you want to delete all the data from your local shared object at once, you can use the clear() method on the SharedObject class. This method purges all data and also deletes the shared object from the local file system of the device. The instance is still active; however, it no longer contains any data.

Load Data from a SharedObject

A local shared object stores its data as key value pairs in a file that is encoded with the ActionScript Message Format. When your `SharedObject` instance is instantiated by calling the `SharedObject.getLocal()` method, the file is decoded, and all the key value pairs are added to the `data` property object of your instance.

Reading the attributes from the `data` property is very similar to writing the data. For example, if you stored a `username` property in the local shared object, the syntax may look something like the following:

`myso.data.username = "jdolce";`

To retrieve the value of the username, simply place the property on the other side of the equals operator:

`var username:String = myso.data.username as String;`

Because the `data` property is a base ActionScript `Object`, it does not know the data types of all its properties and attributes. In order for you to get the best performance out of your ActionScript code, you will want to strongly type all your variables. This has a number of benefits besides the increase in performance. During development, if the compiler knows what data type your variable is, it allows you to use code hinting to see all the properties and methods associated with that class. It also can catch many compiler errors, especially if you are trying to pass the wrong type of object in as an argument to a function.

Using the `as` keyword enables you to cast an unknown variable to a specific data type. If the value that you are casting is not of that type, your variable will be set to `null`.

Load Data from a SharedObject

1. Select the frame on the Timeline to which to add ActionScript code.
2. Open the Actions panel.

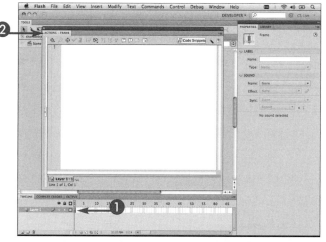

3. Create a `SharedObject` instance, such as `var so:SharedObject = SharedObject.getLocal();`.
4. Enter a name for your `SharedObject`, such as `"scores"`.

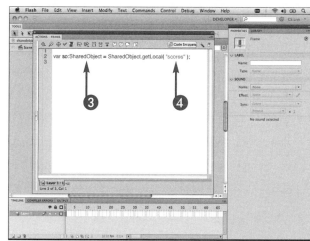

Note: *Save some data in your* `SharedObject`. *For more details, see the preceding section, "Write to a SharedObject."*

5 Retrieve a `String` variable, such as `var username:String = so.data.name as String;`.

6 Retrieve an `int` variable, such as `var score:int = so.data.score as int;`.

7 Trace your data, such as `trace ( username, score );`.

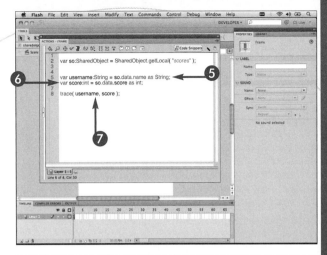

8 Press ⌘+Enter (Ctrl+Enter) to test your movie.

● The data that you retrieved is shown in the Output panel.

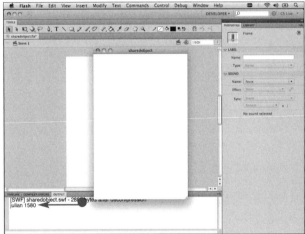

Apply It

If you are unsure of what properties and data you have stored in your local shared object, you can iterate over the `data` property. This enables you to inspect all the key value pairs that you have stored. Using the `for..in` statement allows you to iterate over all the dynamic properties of an object. All fixed properties, such as variables and methods defined in the class definition, are omitted from enumeration. Properties are kept in no particular order and can appear to be random. The order in which your properties appear when iterating over its object will be different every time and should not be relied on. Here is an example:

```
for (var prop:String in so.data )
{
    trace(prop, "=", so.data[ prop ]);
}
```

Connect to a SQLite Database

With the release of Adobe AIR 1.0, we were introduced to a set of ActionScript APIs to communicate with SQLite databases. These allowed developers to store and retrieve information in their applications. iOS also has a SQLite framework, which enables you to communicate with SQLite databases through the same set of ActionScript APIs.

The `SQLConnection` class is used to manage the creation and connection to a SQLite database file. When a `SQLConnection` instance attempts to connect to a database file, it will connect to it if it already exists or create it if it does not.

When you attempt to connect to a database file, you must specify the path in order to successfully connect or create it. You want to carefully pick a location in the application sandbox of your application. Certain directories and folders are not guaranteed to be preserved during an application update. Also, if you have large databases, you should avoid putting them in the Documents folder, as this directory is backed up by iTunes during device sync. Larger database files will cause the backup to take longer. The ideal folder to use the Library/Preferences directory in the user's directory. This will ensure that your database files are maintained when the user updates your application.

When selecting a file for your database file, the entire directory must exist. If one of the folders does not exist, the `SQLConnection` instance will throw an error.

To open the connection to your database file, use the `openAsync()` method on the `SQLConnection` class.

Connect to a SQLite Database

1. Create a `SQLConnection` instance, such as `var conn:SQLConnection = new SQLConnection();`.

2. Create a `File` instance, such as `var db:File = File.userDirectory.resolvePath();`.

3. Type the path to your database file, such as `"Library/Preferences/leaderboard.db"`.

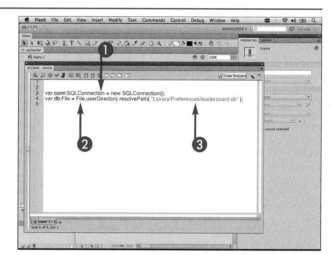

4. Create a listener for when the database is opened, such as `conn.addEventListener(SQLEvent.OPEN, onDatabaseOpen);`.

5. Create a listener for any errors, such as `conn.addEventListener(SQLErrorEvent.ERROR, errorHandler);`.

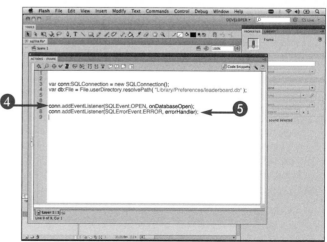

6. Create an event handler for the open event, such as onDatabaseOpen.

7. Add a trace statement to show a successful connection to the database, such as trace("database opened");.

8. Create an event handler for the error event, such as errorHandler.

9. trace any errors that occur, such as trace(event.error.message);.

10. Open the database, such as conn.openAsync(db);.

11. Press ⌘+Enter (Ctrl+Enter) to test your movie.

- Your trace statements appear in the Output panel.

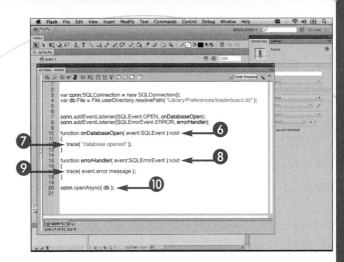

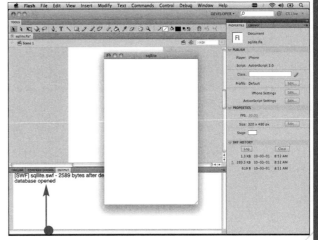

Extra

SQLConnection instances operate in two distinct modes, asynchronous and synchronous. The example in this section uses the openAsync() method, which uses the asynchronous mode. With the asynchronous method, any SQLite database operations occur in the background and separate from the main thread. Each SQLConnection instance runs in its own thread to maximize performance. This allows the user to continue to interact with the application while database operations are executing. When using the asynchronous method, you will need to register event listeners in order to determine when a database operation has successfully completed.

Using the open() method will create a connection using the synchronous execution. This method does not require you to add any event listeners to determine when an operation has successfully completed. This makes the synchronous method more desirable for developers; however, it causes the commands to run in the main thread. This means that any database operation will cause the rest of the application to pause, including any animations and touch events, until the operation is complete. Because events are not fired with this method, you will need to surround your database operation methods with try..catch blocks in order to capture any errors that occur.

Create a SQLite Table

After you have opened a connection to a SQLite database, you can create a table within it to store information. A database table consists of a predetermined number of columns and any number of rows. Each column is identified by its name, which is given upon creation. Along with a unique name, each column has a type of data associated with it.

The following are valid types for your table columns: TEXT, NUMERIC, INTEGER, REAL, Boolean, Date, XML, XMLList, Object, and NONE.

To create a table in your database, you can execute a CREATE TABLE SQL statement. You can also specify to create the table only if it does not already exist. Each table in your database has a unique name in order to distinguish which table you would like to select or insert data into.

The SQLStatement class gives you the ability to execute any SQL statement against a SQLConnection instance that has already been opened. Setting the sqlConnection property of the SQLStatement class allows you to link your SQL statement to an open SQL connection. The text property of the SQLStatement class specifies the actual SQL statement to execute, in this case the CREATE TABLE statement.

After your SQL statement is created and linked to an open connection, you can call the execute() method in order to execute the SQL statement on your database. If your SQLConnection instance was opened using the asynchronous method, you will need to register event listeners on your SQLStatement instance in order to determine when your statement has finished executing successfully. If you are using the synchronous method, be sure to surround your execute code with a try..catch block to catch any errors.

Create a SQLite Table

Note: The code shown here continues the example from the section "Connect to a SQLite Database."

1. Create a String instance variable, such as `var sql:String = "";`.

2. Create a table if it does not exist, such as CREATE TABLE IF NOT EXISTS.

3. Give your table a name, such as scores.

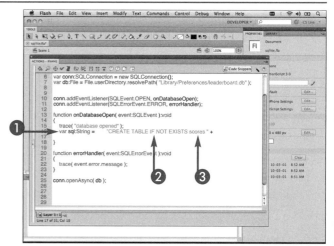

4. Create an ID column, such as `"( id INTEGER PRIMARY KEY AUTOINCREMENT, " +`.

5. Create a name column, such as `"name TEXT, " +`.

6. Create a score column, such as `"score INTEGER )";`.

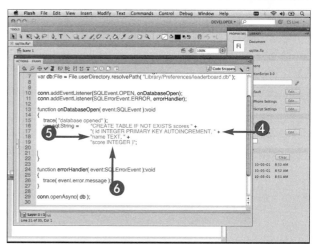

7 Create a SQLStatement variable, such as `var sqlStat:SQLStatement = new SQLStatement();`.

8 Set the connection for the SQL statement, such as `sqlStat.sqlConnection = conn;`.

9 Set the SQL string to the statement, such as `sqlStat.text = sql;`.

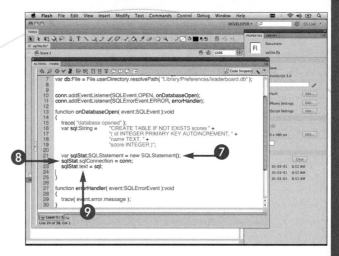

10 Add a listener for the successful execution of the statement, such as `sqlStat.addEventListener(SQLEvent.RESULT, tableResult);`.

11 Add a listener for any SQL errors, such as `sqlStat.addEventListener(SQLErrorEvent.ERROR, errorHandler);`.

12 Execute the SQL statement, such as `sqlStat.execute();`.

13 Create an event handler for the result, such as `tableResult`.

A table is created in the SQLite database.

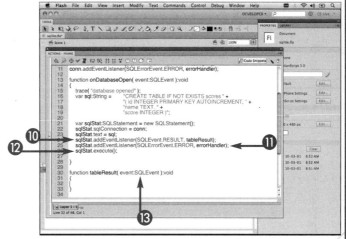

Apply It

For each column created in your database table, you can specify a default value when doing an INSERT command. This can be done by using the DEFAULT constraint after the column data type, and the value can be NULL, a string constant, a number, or an expression enclosed in parentheses. This is especially useful when you want to add the time in which the row was inserted into the database into a date or time column. To help with this, there are three special keywords: CURRENT_TIME, CURRENT_DATE, and CURRENT_TIMESTAMP. The CURRENT_TIME keyword format is HH:MM:SS, the CURRENT_DATE format is YYYY-MM-DD, and the CURRENT_TIMESTAMP format is YYYY-MM-DD HH:MM:SS. Here is what the SQL statement would look like to add a time column with a default value to the example's scores table:

```
var sql:String = "CREATE TABLE IF NOT EXISTS scores (" +
" id INTEGER PRIMARY KEY AUTOINCREMENT, " +
" name TEXT, " +
" score INTEGER," +
" time DATE DEFAULT CURRENT_TIMESTAMP" +
")";
```

Insert Data into a SQLite Table

Inserting data into a SQLite database table is similar to creating the table. You use the same methods for creating and executing a SQLStatement instance as you did earlier. The only difference is the text property, which represents the SQL statement to execute. The INSERT SQL command is used to insert values into the table.

The INSERT statement uses the VALUES keyword to insert a single row into an existing table. If no column list is specified, the number of values specified must match the same number of columns in the table. If a column list is present, the number of values must match the number of columns specified. Any columns in the table that are not specified in the column list will be filled with their default value or with NULL if no default value is specified.

You should be aware that there are security concerns when inserting values when a user has the ability to input data. Concatenating an INSERT statement with user input could allow the user to enter another SQL statement, which could be run and affect your database — potentially erasing it all. This is called a *SQL injection attack* and can be prevented with the use of the parameters property on the SQLStatement class.

Parameters are used to allow for the typed substitution of unknown values at the time of the SQL construction. The use of parameters is the only way to guarantee that the proper data type will be stored in the database. If parameters are not used, the text representation of the class based on the associated column's type affinity will be used.

Insert Data into a SQLite Table

Note: *The code shown here continues the example from the section "Create a SQLite Table."*

1. Create a String variable, such as var sql:String = "";.

2. Insert data into your table, such as INSERT INTO scores.

3. List the columns to insert data into, such as (names,score).

4. List parameters for the values, such as values(:name, :score).

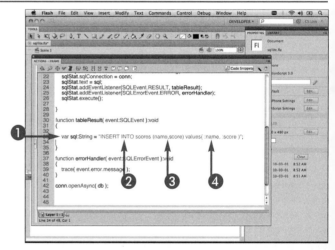

5. Create a SQLStatement variable, such as var sqlStat: SQLStatement = new SQLStatement();.

6. Set the connection for the SQL statement, such as sqlStat.sqlConnection = conn;.

7. Set the SQL string to the statement, such as sqlStat.text = sql;.

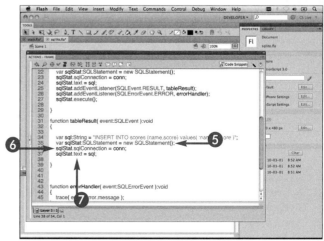

8 Set the parameter for the name value, such as `sqlStat.parameters[ ":name" ] = "julian";`.

9 Set the parameter for the score value, such as `sqlStat.parameters[ ":score" ] = 150;`.

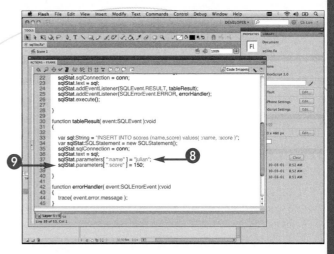

10 Add a listener for the successful execution of the statement, such as `sqlStat.addEventListener(SQLEvent.RESULT, onInsert);`.

11 Add a listener for any SQL errors, such as `sqlStat.addEventListener(SQLErrorEvent.ERROR, errorHandler);`.

12 Execute the SQL statement, such as `sqlStat.execute();`.

13 Create an event handler for the result, such as `onInsert`.

A row is inserted into the table in the database.

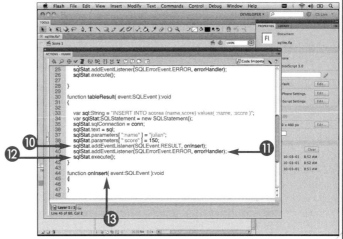

Extra

The example here explores one of the three different techniques to use the INSERT statement. The second INSERT statement method takes its data from a SELECT statement. If no column list is specified, the number of columns in the result of the SELECT statement must exactly match the number of columns in the table. If a column list is specified, the number of columns in the result must match the number named in the list. A new entry is created for every row returned by the SELECT statement. For more details on how to use the SELECT statement, see the following section, "Select Data from a SQLite Table."

For example, you could have created a table in your database called `topfive`, in which you want to insert the top five scores from your scores table. Your SQL statement would look like this:

```
var sql:String = "INSERT INTO topfive (name,score) SELECT name,score FROM scores ORDER BY score
   DESC LIMIT 5";
```

The third INSERT statement method uses the default values of the columns in the table. A new row is created in the database with each column filled with its default value.

Select Data from a SQLite Table

After your SQLite database table is populated with data, you can use the SELECT SQL statement to retrieve specific data from it. Retrieving data from a table is a two-step process. The first step is to execute a SQL SELECT statement with a SQLStatement instance.

SELECT statements can be very simple or very complex. In its simplest form, the SELECT statement enables you to select specified columns from a table in the database. You can also use the * in place of specific columns to select all the columns of the table. This method will return all the data for each row returned by the SQL statement.

You can use the WHERE clause in order to limit the number of rows that are returned. An equation is used after the WHERE keyword in order to filter which rows to return, such as SELECT * FROM scores WHERE name='julian'.

You can also use the LIMIT clause to place an upper bound on the number of results to return. A negative number will cause there to be no upper bound.

After the SELECT statement has successfully executed and returned results from the database, you can retrieve a SQLResults object by calling the getResults() method on the same SQLStatement instance that executed the SELECT. The data property of the SQLResults object is an Array containing all the results of the SQL statement that was executed. Each entry in the Array represents a row in the database table. The data property will be null if the SQL statement returns 0 rows or if it is not a SELECT statement.

Select Data from a SQLite Table

Note: The code shown here continues the example from the section "Insert Data into a SQLite Table."

1. Create a String variable, such as `var sql:String = "";`.

2. Select what columns to select from the table, such as SELECT *.

3. Type **FROM**.

4. Select what table to select from, such as scores.

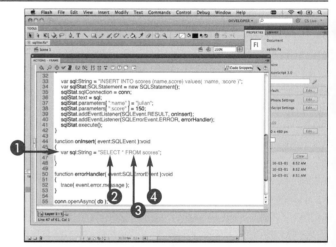

5. Create a SQLStatement variable, such as `var sqlStat:SQLStatement = new SQLStatement();`.

6. Set the connection for the SQL statement, such as `sqlStat.sqlConnection = conn;`.

7. Set the SQL string to the statement, such as `sqlStat.text = sql;`.

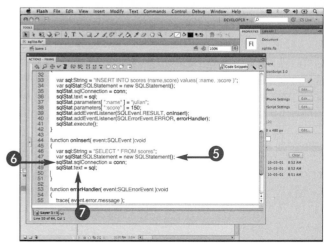

8. Add a listener for the successful execution of the statement, such as `sqlStat.addEventListener(SQLEvent.RESULT, onSelect);`.

9. Add a listener for any SQL errors, such as `sqlStat.addEventListener(SQLErrorEvent.ERROR, errorHandler);`.

10. Execute the SQL statement, such as `sqlStat.execute();`.

11. Create an event handler for the result, such as `onSelect`.

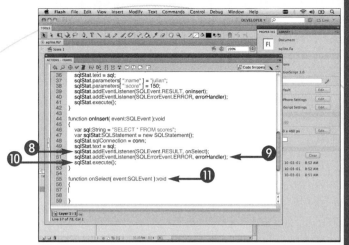

12. Create a SQLResult variable, such as `var result:SQLResult = SQLStatement( event.target ).getResult();`.

13. Loop through the rows.

14. Iterate over the properties of the row.

15. Output the property and its value, such as `trace( prop, result.data[ i ][prop] );`.

    The queried data is returned from the database.

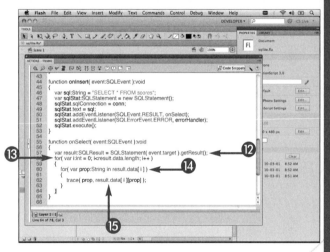

Apply It

When using the SELECT statement to return a set of rows from the database, you may want to have them ordered in a certain way. By default, they are returned in the same order that they were inserted into the database. If you take the leaderboard example, you will want to return the scores in order from highest to lowest. Ordering the results as they are returned is far more efficient than ordering them after they have been returned in ActionScript. You can use the ORDER BY clause to order the returned results in conjunction with your SELECT statement:

```
var sqlStat:SQLStatement = new SQLStatement();
sqlStat.sqlConnection = conn;
var sql:String = "SELECT * FROM scores ORDER BY score DESC";
sqlStat.text = sql;
sqlStat.addEventListener(SQLEvent.RESULT, onSelect);
sqlStat.addEventListener(SQLErrorEvent.ERROR, createError);
sqlStat.execute();
```

Update Data in a SQLite Table

After your database table becomes populated with data, you may find yourself wanting to update the data in the table. One option would be to delete all the rows that you would like to update and then reinsert them into the table with their new values. This would require more steps than necessary and would not be very efficient. There is a SQL command that allows you to do this all in one step. The UPDATE SQL statement allows you to update the value of a specific column in selected rows of a table.

The SET keyword is used in order to specify specific columns in your table and the new values you would like to update them to. Each assignment in an UPDATE statement specifies a column name to the left of the equals sign and an arbitrary expression to the right of it. The expression can be a value from other columns or a completely new one. It is important to note that all expressions are evaluated before any assignments are made. To update multiple columns at the same time, separate each assignment with a comma.

A WHERE clause can also be used in your UPDATE statement to restrict which rows in the table are updated. Omitting the WHERE clause will cause all rows to be affected by the update. The ORDER BY clause in an UPDATE statement is used only to determine which rows fall within the limit. The order in which the rows are modified is arbitrary and is not determined by the ORDER BY clause.

Update Data in a SQLite Table

Note: *The code shown here continues the example from the section "Select Data from a SQLite Table."*

1. Create a `String` variable, such as `var sql:String = "";`.
2. Type **UPDATE**.
3. Set the database table name, such as `scores`.
4. Type **SET**.

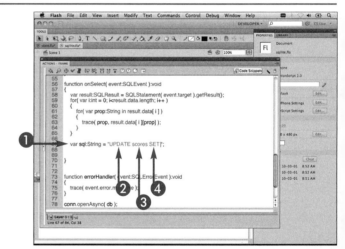

5. Select a column to update, such as `name=`.
6. Set the new value of the column, such as `'Julian Dolce'`.
7. Type **WHERE**.
8. Select a column to check against, such as `name=`.
9. Set the value to check against, such as `'julian'`.

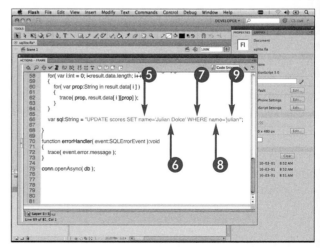

188

⑩ Create a SQLStatement variable, such as var sqlStat:SQLStatement = new SQLStatement();.

⑪ Set the connection for the SQL statement, such as sqlStat.sqlConnection = conn;.

⑫ Set the SQL string to the statement, such as sqlStat.text = sql;.

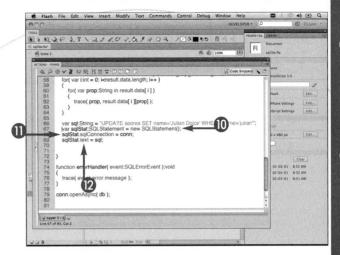

⑬ Add a listener for the successful execution of the statement, such as sqlStat.addEventListener(SQLEvent.RESULT, onUpdate);.

⑭ Add a listener for any SQL errors, such as sqlStat.addEventListener(SQLErrorEvent.ERROR, errorHandler);.

⑮ Execute the SQL statement, such as sqlStat.execute();.

⑯ Create an event handler for the result, such as onUpdate.

The specified rows in the database are updated with new values.

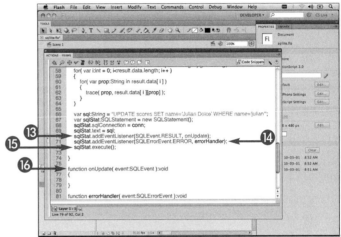

Extra

You can make specific columns in your database table contain unique values. In the example here, you could have specified the name column to be unique. This would mean that when a user tried to submit a second score with the same name, it would be ignored, as there is already a row with the same value. To work around this, you would check to see if the name already exists and then update the score value and if it did not exist, insert the data into the database.

You can also use the INSERT OR REPLACE statement, which will do this for you in one statement. The following SQL statement syntax checks the name value for any rows that already exist with the same value and either updates or inserts the new row:

```
var sql:String = "INSERT OR REPLACE INTO scores (name,score) values ('julian', 5000)";
```

Delete Data from a SQLite Table

You are able to delete rows from your database using the DELETE SQL statement. The DELETE statement starts with the DELETE FROM keywords and is followed by the name of the table from which it will remove records. Following the table name is the WHERE keyword, which can be used to supply an expression to evaluate against. Those rows that match the expression will be removed. The result of the expression must always return either true or false in order to properly determine whether it should remove the row or not.

If no WHERE keyword and expression are supplied, all rows will be removed from the table. When using this method to delete all the rows from a table, SQLite uses an optimization in order to remove all the rows, without having to visit each row of the table individually. This truncate optimization makes the delete run much faster.

When using string values in your expressions, you must use single quotes around the value that you want to evaluate against. For example, WHERE name=julian will throw an error because it tries to look for the column named julian. To fix this, here is the proper syntax: WHERE name='julian'. This applies only to strings, as integer values do not require quotes in order to properly evaluate.

You can also combine multiple expressions to narrow down the set of rows to delete from the table. To combine the expressions, you can use the AND keyword. For example, the following SQL statement would remove all rows in the table in which the name is equal to "julian" and the score is equal to 4000: DELETE FROM scores WHERE name='julian' AND score=4000.

Delete Data from a SQLite Table

Note: The code shown here continues the example from the section "Update Data in a SQLite Table."

① Create a function that will delete names from the table, such as `deleteName`.

② Pass in a `name` argument as a `String`, such as `name:String`.

③ Create a `String` variable, such as `var sql:String = "";`.

④ Type **DELETE FROM**.

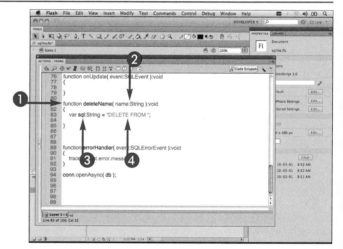

⑤ Set the table name, such as `scores`.

⑥ Type **WHERE**.

⑦ Set the column to check against, such as `name=`.

⑧ Create a parameter index, such as `:name`.

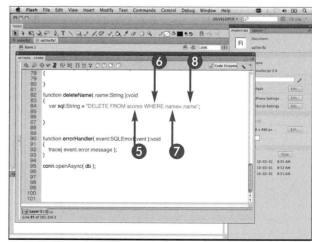

9 Create a SQLStatement variable, such as `var sqlStat:SQLStatement = new SQLStatement();`.

10 Set the connection for the SQL statement, such as `sqlStat.sqlConnection = conn;`.

11 Set the SQL string to the statement, such as `sqlStat.text = sql;`.

12 Set the parameter for the name that is passed in, such as `sqlStat.parameters[ ":name" ] = name;`.

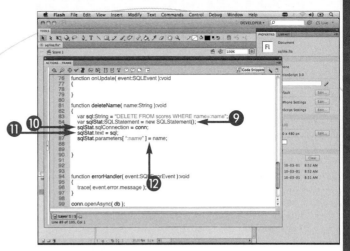

13 Add a listener for the successful execution of the statement, such as `sqlStat.addEventListener(SQLEvent.RESULT, onDelete);`.

14 Add a listener for any SQL errors, such as `sqlStat.addEventListener(SQLErrorEvent.ERROR, errorHandler);`.

15 Execute the SQL statement, such as `sqlStat.execute();`.

16 Create an event handler for the result, such as `onDelete`.

The specified rows are deleted from the table in the database.

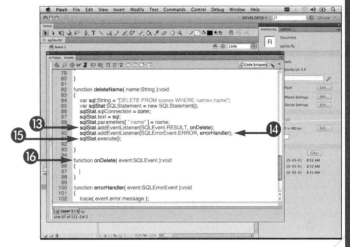

Extra

Indexes help us retrieve information quicker. For example, if you are looking for a specific topic in this book, it is quicker to look it up in the index, instead of reading the book from the start in order to find it. The same applies for tables in your database. When selecting data using the SELECT statement, every row is visited, which could affect speed, depending on how many rows you have in the table. Creating an index for specific columns will reduce the number of rows that need to be visited when running your SELECT statement. In the leaderboard example, you may want to create an index on the name column if you are doing a lot of selects based on its value. The trade-off, however, is an index will cause INSERT statements to be slower because the index needs to be regenerated after every insert. You could use the DROP INDEX statement in order to delete the index before an INSERT and then re-create it afterwards. In most instances, this is necessary only when dealing with very large data sets. To create your index, execute the following SQL statement:

```
var sql:String = "CREATE INDEX name_idx ON scores (name)";
```

Handle Application Exits

Currently, iOS can have only one third-party application running at a time. This means that after your application has exited, it is no longer running and cannot run as a background task. Your application can exit in several different ways. When a user presses the home button on the device, it causes the currently running application to quit and exit to the home screen. The user can also click a link inside your application that launches another application, such as the YouTube or Mail application. Your users can also be interrupted by a phone call, SMS, or push notification alert. Accepting these interruptions will cause your application to exit and open the appropriate application.

The NativeApplication class dispatches an event when your application is exiting. Listening for this event allows you to perform any needed last-second cleanup of objects or save the state of your application. When listening to the Event.EXITING event, it is a good idea to remove the listener after it has been fired once. There are instances in which it may be fired twice, such as when answering a phone call, and executing your exit code may result in unwanted errors.

You may have noticed at times when exiting an application when sound is playing that it often continues to play for a few seconds after the application has exited. This is caused by the application not being fully removed from memory when it exits and can cause some undesired results for your user. It is a good practice to stop all the sounds in your application when your application quits. Calling the SoundMixer.stopAll() method will stop all currently playing sounds.

Handle Application Exits

1. Create a NativeApplication variable, such as `var application :NativeApplication;`.

2. Set the variable to your application's NativeApplication, such as `= NativeApplication. nativeApplication.`.

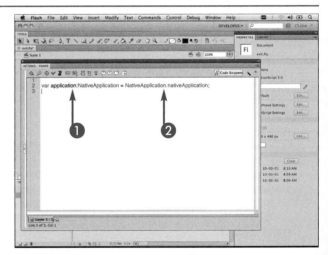

3. Add a listener to your application, such as `application.addEventListener( );`.

4. Listen for the Event.EXITING event.

5. Set an event handler to be fired, such as onExit.

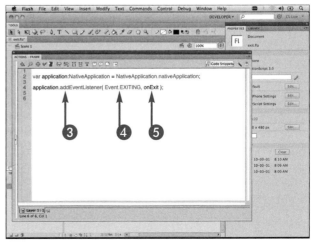

6 Create an event handler function for the event, such as `onExit`.

7 Output that the event has fired, such as `trace( "exiting" );`.

8 Stop all sounds, such as `SoundMixer.stopAll();`.

9 Remove the event listener, such as `application.removeEventListener( Event.EXITING, onExit );`.

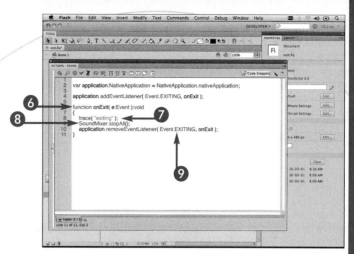

10 Press ⌘+Enter (Ctrl+Enter) to test your movie.

11 Close your application.

12 Watch the Output panel to see the event get fired.

The `Event.EXITING` event is fired when the application is closed.

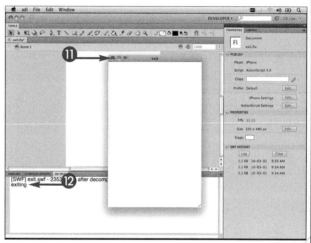

Extra

When your application exits, you will want to save any necessary data as fast as possible. If the save process takes too long, there is a chance that the application will quit without completing the entire save. In order to prevent this, you will want to limit the data to be as small as possible.

After all your data is saved, you can try and do any cleanup of your application. This would be no different than the regular code cleanup you would do when disposing of objects that are no longer needed. Make sure that all your animations and Timelines have stopped. Stop all timers and enter frame loops. Remove any remaining listeners on any live objects. Also, close any network requests, sockets, and file streams that you may still have open.

All of these will make sure that your application is completely cleared from memory when your application quits. If your application causes memory leaks when it closes, the user may have to reboot his or her device in order to clear its memory.

Save Application States

If you use the Mail application often, you have probably noticed that the application starts in the same state as when you last exited it. Because only one third-party application can be running at any given time on iOS, Apple strongly recommends that your application save its state often. And when your application is relaunched, launch it in the state that it was last saved in.

Another reason it is recommended you save the state of your application often is that your application can quit unexpectedly. As mentioned previously, listening for the application exiting event is a great place to save the state of your application. However, you should not rely on this too much, as that event may not get fired every time your application quits. This is especially true in the event that your application crashes or uses too much memory.

It will be up to you to determine how often you save the state of your application. Different types of applications will call for different methods. For example, if you are developing a game, you may want to save more often than an application.

This chapter has discussed two different methods of how to save states of your application. A SharedObject is probably all that is needed when saving states as a SQLite database may be overkill. Saving the state information of your application to a file on the device is an additional option. For more details on working with files, see Chapter 11, "Working with Files." You will want to implement the method that will allow you to not only save the data efficiently, but also load it quickly when your application starts.

Save Application States

1. Create a function to save the section when it changes, such as `function saveCurrentSection():void{}`.

2. Set a `String` argument as the section to save, such as `section:String`.

3. Create a `SharedObject` variable, such as `var so:SharedObject = SharedObject.getLocal();`.

4. Set the name of your `SharedObject`, such as `"state"`.

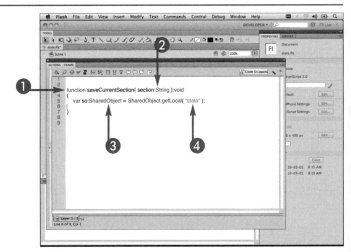

5. Save the section in the `SharedObject`, such as `so.data.section = section;`.

6. Save the data to the disk, such as `so.flush();`.

Note: *Call this function every time your application changes its section.*

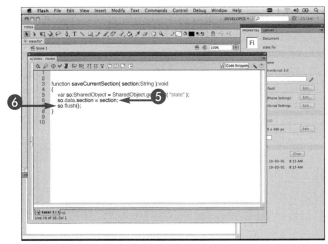

194

Chapter 10: Saving State

7 Create a function to load the last saved section, such as `function getLastSection()`.

8 Give it a return type of `String`.

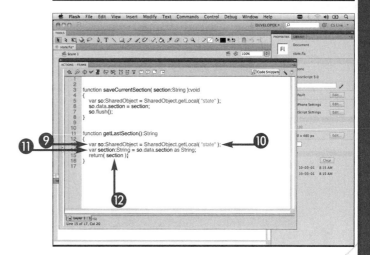

9 Create a `SharedObject` variable, such as `var so:SharedObject = SharedObject.getLocal();`.

10 Enter the name of your `SharedObject`, such as `"state"`.

11 Get the section from the `SharedObject`, such as `var section:String = so.data.section as String;`.

12 Return the section, such as `return ( section );`.

The current section is saved to a local `SharedObject`.

Extra

Saving the state of your application does not have to be a complex process. You do not have to save every single interaction point between the user and your application. Simply saving which section of your application was last viewed is usually enough for most applications. But there are instances in which the data may become more complex, and it will be up to you to decide what data is most important to save for when the user returns. In the instance in which you are creating a game, you should save the state after each level at the very least. This allows the user to continue the game, also saving his or her score. Depending on the type of game, however, you may be able to save more frequently, such as when the user takes a turn or makes a move in the game. This can easily be done with a casual game but is harder for an action or driving game. When deciding what data and how often to save it, ask yourself what makes sense for the user to continue when the game is loaded. You can also give them an option upon relaunching the game to continue the last game or start a new one.

195

Reference Files and Directories

An iPhone application has access to read and write files to the local file system of the device. Each application is placed in its own directory and, for security purposes, has access to read and write to those directories only. The path to your application on the device is similar to /var/mobile/Applications/AF22434E-C0F2-42EA-BC4F-BD1D9DCB10AF/. This path is known as the *application home* and is often referred to by Apple as <Application_Home>.

There are a few common directories that are created in the application home directory that you can use to write preference files, save documents, and save the state of your application. The File class contains static properties that allow you to access these directories.

File.applicationDirectory contains your application and bundled files. Because each application must be signed, you cannot write files to this directory. However, you can read files from it.

File.applicationStorageDirectory is located at <Application_Home>/Library/Application Support/<app ID>/Local Store. This is the same directory that your SharedObjects are saved in.

File.documentsDirectory is the Documents directory that you should use to store any application-specific data or any information that should be backed up regularly. The contents of this directory are backed up by iTunes.

File.userDirectory gives you the path to the application home directory. This allows you to access the <Application_Home>/Library/Preferences directory, which is where any preference files are stored.

Reference Files and Directories

Create Text Fields

1. Click the Text tool.
2. Draw a text field on the Stage.
3. Click here and select Classic Text.
4. Click here and select Dynamic Text.
5. Give the text field an instance name, such as application_txt.
6. Click here and select Multiline.
7. Click here and select a device font, such as Helvetica.
8. Click here and select Use Device Fonts.
9. Copy the text field on the Stage.
10. Paste a new text field on the Stage.
11. Give it an instance name, such as documents_txt.

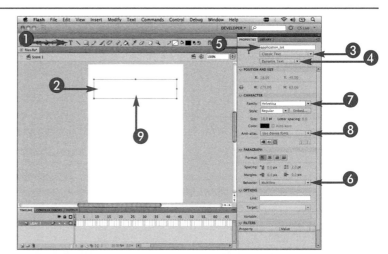

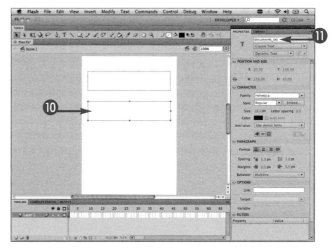

12 Paste another new text field on the Stage.

13 Give it an instance name, such as `user_txt`.

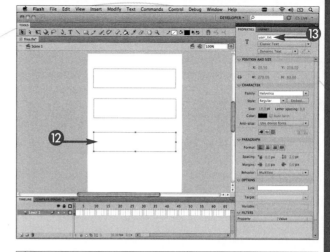

Set the Text Fields to Your Application's Directories

14 Open the Actions panel.

15 Set the text to the application path, such as `application_txt.text = File.applicationDirectory.nativePath;`.

16 Set the text to the documents path, such as `documents_txt.text = File.documentsDirectory.nativePath;`.

17 Set the text to the application home path, such as `user_txt.text = File.userDirectory.nativePath;`.

The path to the application directories are placed in the text fields on the Stage.

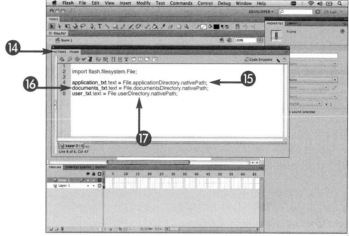

Extra

When your application is backed up by iTunes, it performs an incremental backup of all files, except for those in the following directories: File.applicationDirectory, <Application_Home>/Library/Caches, and <Application_Home>/tmp. To make sure that your application does not take a long time to back up, you should be selective on where you place your files. The Documents directory should store only files that the application cannot easily re-create. The bigger the files in this directory, the longer it will take for your backup to complete, which can be annoying for your users. Instead, store files in the <Application_Home>/Library/Caches directory. You can also use the <Application_Home>/tmp directory to store any temporary files that your application may create.

When the user updates your application, a new application directory is created, and the application is installed. All the user's application data is then moved to the new application directory, and then the old installation is deleted. Files in your Documents directory and <Application_Home>/Library/Preferences directory are guaranteed to be preserved and copied to the new location. Other directories may be copied over as well, but there is no guarantee, so you should not rely on their being present after an update.

Write Files

You can write files to the local file system of the device using the `File` class and the `FileStream` class. You can write any type of file to the file system as long as you know how to properly encode the file. Any text file format is the easiest type of file to write to the file system.

The first step to writing a file is to create a `File` instance, which is a representation of a path to a file or directory on the file system. This can be an existing file or directory or one that does not currently exist. The `File` class has a number of static properties that map to specific directories on the file system, as shown in the section "Reference Files and Directories." You can use the `resolvePath()` method to create a new `File` object with a path relative to the current path of the `File` instance.

After you have created a reference to your file location, you can use the `FileStream` class to write to the file. The `FileStream` class is very similar to the `ByteArray` class and has all the same methods for reading and writing bytes to the object. In most cases, you will simply be dealing with text files and can use the `FileStream.writeUTFBytes()` method in order to write any `String` instance variables to the file.

After you have written all the data to the file, make sure to close the stream by calling the `FileStream.close()` method. Upon closing the stream, you will no longer be able to write data to the file.

Write Files

1. Create a `File` variable, such as `var file:File`.

2. Select a directory for your file, such as `File.documentsDirectory`.

3. Resolve the path, such as `resolvePath();`.

4. Create a name for your file, such as `preferences.txt`.

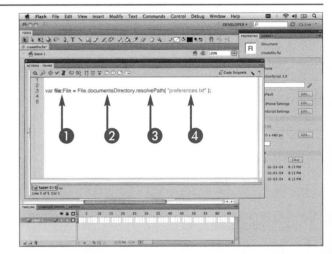

5. Create a `FileStream` variable, such as `var stream:FileStream = new FileStream();`.

6. Add a listener to the stream, such as `stream.addEventListener();`.

7. Listen for the `IOErrorEvent.IO_ERROR` event.

8. Specify an event handler function, such as `onIOError`.

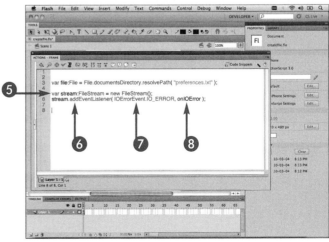

9 Create the event handler function, such as `onIOError`.

10 Output the event, such as `trace( event );`.

11 Open the stream, such as `stream.openAsync();`.

12 Select the file to open.

13 Open the file in the `FileMode.WRITE` mode.

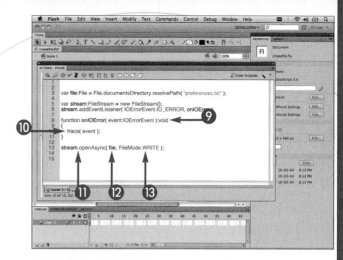

14 Write UTF bytes to the stream, such as `stream.writeUTFBytes();`.

15 Enter a URL-encoded string containing name/value pairs, such as `"username=julian&password=mypassword"`.

16 Close the stream, such as `stream.close();`.

A file is created with the specified text on the device.

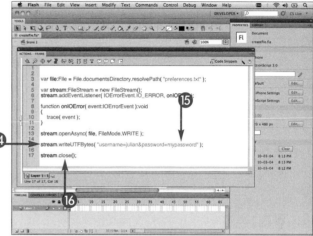

Apply It

You can also write a `ByteArray` instance to the `FileStream` object. This makes it very easy to save nontext-based file formats such as images. Chapter 6, "Working with Images," discusses encoding PNG images from a `BitmapData` instance. The PNG image is encoded into a `ByteArray`, which can be saved to the file system:

```
var bd:BitmapData = new BitmapData( 200, 200, false, 0xFFFF0000 );
var png:ByteArray = PNGEncoder.encode( bd );
var stream:FileStream = new FileStream();
var file:File = File.documentsDirectory.resolvePath( "my.png" );
stream.open(file, FileMode.WRITE );
stream.writeBytes( png );
stream.close();
```

Read Files

Reading files is very similar to writing files. You can read any file that you have access to on the file system. To select a file to read, you create a `File` object to the path of the file that you want on the file system.

When reading files, it is always a good idea to make sure that it exists before trying to read it. The `exists` property on the `File` class will return `false` if it does not exist and `true` if it does.

If your file exists, you can create a `FileStream` instance that points to your `File` and use the `FileMode.READ` property when calling the `openAsync()` method. Opening the file asynchronously causes the file to be opened and read into the buffer on a separate thread. This prevents any performance degradation on the main thread, which may cause animations and sound to pause while reading the file.

In order to determine when the file has been completely read into the buffer, you can listen for the `Event.COMPLETE` event on the `FileStream` object. After the file has been completely read into the buffer of the `FileStream` instance, you can use any of the `read` methods on it to read the data. When you are reading any kind of text file, you can use the `readUTFBytes()` method. When reading UTF bytes, you need to specify how many bytes you would like to read at any given time. You can use the `bytesAvailable` property on the `FileStream` class to determine how many bytes there are from the current position of the buffer to the end of the buffer. If the current position of the buffer is 0, this will allow you to read the entire file.

Read Files

1. Create a `File` variable, such as `var file:File`.

2. Select a directory for your file, such as `File.documentsDirectory`.

3. Resolve the path, such as `resolvePath();`.

4. Create a name for your file, such as `preferences.txt`.

5. Create a `FileStream` variable, such as `var stream:FileStream = new FileStream();`.

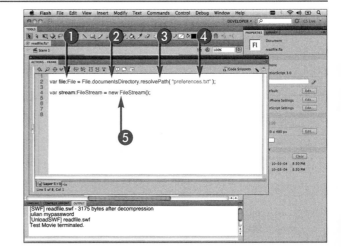

6. Listen for the complete event on the stream, such as `stream.addEventListener( Event.COMPLETE, onComplete );`.

7. Open the file, such as `stream.openAsync();`.

8. Select the file to open.

9. Open the file in the `FileMode.READ` mode.

10. Create an event handler method, such as `onComplete`.

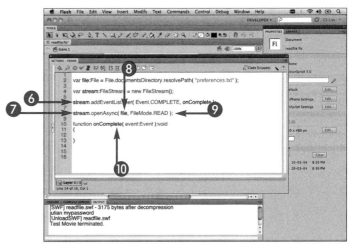

⑪ Create a String variable, such as `var str:String`.

⑫ Read UTF bytes of the stream, such as `stream.readUTFBytes();`.

⑬ Read all the bytes available in the stream, such as `stream.bytesAvailable`.

⑭ Create a URLVariable variable, such as `var variables:URLVariables = new URLVariables();`.

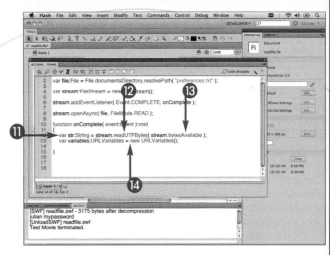

⑮ Decode the text read from the file, such as `variables.decode( str );`.

⑯ Output your variables, such as `trace( variables.username, variables.password );`.

⑰ Close the stream, such as `stream.close();`.

The text is read from the file.

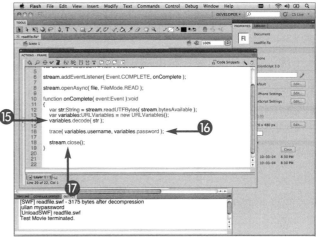

Apply It

You can also read an entire file to a `ByteArray` instance. This is very useful when dealing with binary file formats, such as images, databases, and binary property list files. You will need to understand the format of the file in order to properly parse a binary file. After you have your file in a `ByteArray` instance, you can also send it over a `Socket` connection. Connecting to a socket server on your computer or over the network enables you to send your files to your computer to back them up and save them:

```
var ba:ByteArray = new ByteArray();
stream.readBytes( ba );
//sending the files over a connected Socket class
ba.position = 0;
socket.writeBytes( ba );
socket.flush();
```

Update Files

When you open a file in either the `FileMode.READ` or `FileMode.WRITE` mode, you can perform only the specified task. For example, when reading a file, you can only read the file and cannot use any of the write methods to write data to the file. The same is true for when you are writing a file.

The `FileMode.UPDATE` mode allows you to open a file that you can read and write at the same time. This comes in handy when dealing with preferences files. For example, when your application starts up, it will read a preferences file from the file system in order to initialize your application. This could be something simple such as automatically logging in the user or determining which section of your application to start in. After the file has been read, there is a high probability that you will want to overwrite the data in the file to reflect the user's interactions with your application.

Without the ability to open the file in the update mode, you would have to open the file twice — once to read it and once to write it. Opening the file in the update mode enables you to keep the stream to the file open for as long as you need. This can have a dramatic effect on performance as you do not have to open the file and re-create a new `FileStream` and `File` instance.

It is important to keep track of the position of the buffer as you read and write to it. When you initially read the entire file, the position of the buffer is located at the end. If you want to overwrite the entire file, you will need to set the position of the buffer to the beginning of the file. Otherwise, you will simply write your data at the end of the file.

Update Files

1. Create a `File` variable, such as `var file:File;`.

2. Select a directory for your file, such as `File.documentsDirectory.resolvePath()`.

3. Create a name for your file, such as `preferences.txt`.

4. Create a `FileStream` variable, such as `var stream:FileStream = new FileStream();`.

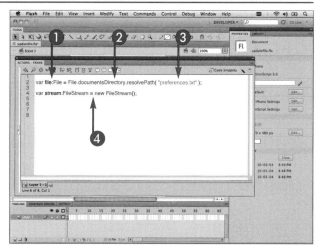

5. Listen for the complete event on the stream, such as `stream.addEventListener( Event.COMPLETE, onComplete );`.

6. Open the file, such as `stream.openAsync();`.

7. Select the file to open.

8. Open the file in the `FileMode.UPDATE` mode.

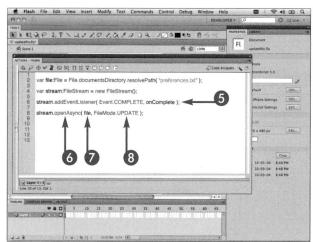

9. Create an event handler method, such as onComplete.

10. Create a String variable, such as var str:String.

11. Read the entire file, such as stream.readUTFBytes(stream.bytesAvailable);.

Note: The position of the buffer in the stream is now located at the end of the file.

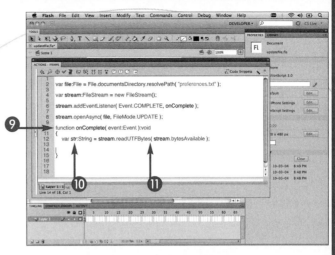

12. Write UTF bytes to the stream, such as stream.writeUTFBytes();.

13. Create a new parameter, such as "&score=1234".

14. Close the stream, such as stream.close();.

The file is updated with the new data.

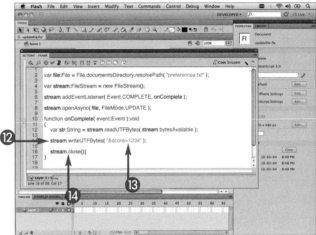

Apply It

The truncate() method on the FileStream class enables you to trim the file from its current location specified by the position property. This allows you to keep your files below a given file size easily. For example, you may have set up a logging system in your application to log events or errors to a file on the file system of the device. Over time, this file could grow to be unnecessarily large. You can use the File.size property in order to determine the size of the file in bytes. The following is an example of truncating a file to keep it under a given size:

```
var file:File = File.documentsDirectory.resolvePath("logs.txt");
var stream:FileStream = new FileStream();
stream.open(file, FileMode.UPDATE);
if (file.size > 1000)
{
    stream.position = 1000;
    stream.truncate();
}
stream.close();
```

Append Files

If you know that you always want to write to the end of the file, you can use the `FileMode.APPEND` mode when opening your file. When the file is opened in the append mode, data is always written to the end of the file when any write method is called on the `FileStream` instance. This makes it easier to write to the end of the file, as you never have to worry about what position the buffer is currently at before writing the data.

If the file does not exist prior to opening it, the file will still be created just as it would when opening the file in the write mode. Also, when the file is opened in append mode, you are able only to write to the file and cannot read it. If you need the ability to read and write the file, open the file with the `FileMode.UPDATE` mode. For more details on how to update files, see the preceding section, "Update Files."

Popular uses for opening files in the append mode is when you are writing to a log file. A log file can come in a variety of forms, such as a history of the user's actions when interacting with your application or a list of errors if any are to occur in your application. These can be valuable, especially when you allow other people to test your application during development. If the user experiences a crash or finds a portion of the application that is broken, you can have him or her send you the log files saved by the application. This will give you more details and help you track down the problem faster and more effectively.

Append Files

1. Create a `File` variable, such as `var file:File;`.

2. Select a directory for your file, such as `File.documentsDirectory.resolvePath()`.

3. Create a name for your file, such as `preferences.txt`.

4. Create a `FileStream` variable, such as `var stream:FileStream = new FileStream();`.

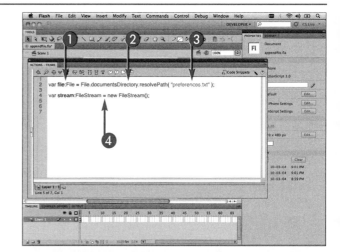

5. Add a listener to the stream, such as `stream.addEventListener();`.

6. Listen for the `IOErrorEvent.IO_ERROR` event.

7. Specify an event handler function, such as `onIOError`.

8. Create the event handler method.

9. Output the event, such as `trace( event );`.

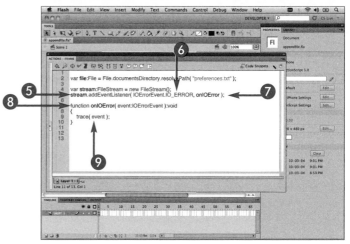

⑩ Open the file, such as `stream.openAsync();`.

⑪ Select the file to open.

⑫ Open the file in the `FileMode.APPEND` mode.

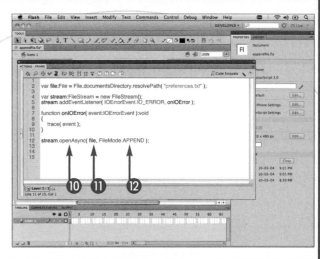

⑬ Write text to the end of the file, such as `stream.writeUTFBytes();`.

⑭ Create a new value to save, such as `"&phone=555-555-5555"`.

⑮ Close the stream, such as `stream.close();`.

The specified data is appended to the end of the file.

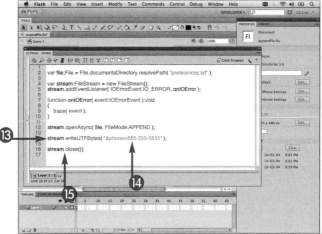

Apply It

The new network API classes in AIR 2.0 make it easy to create an application for the desktop that can communicate with your iPhone application. The new `ServerSocket` class enables you to accept incoming `Socket` connections from the device. After the data is received from a connected socket, you can save the file to the local file system. Here is a very basic example of an AIR application that creates a socket server:

```
var server:ServerSocket = new ServerSocket();
server.addEventListener(ServerSocketConnectEvent.CONNECT, onConnect );
server.bind( 5555, "localhost" );
server.listen();
private function onConnect( event:ServerSocketConnectEvent ):void{
var socket:Socket = event.socket;
//listen for any data arriving on this socket.
}
```

Handle Files Synchronously

There are two ways to open files, asynchronously and synchronously. Throughout this chapter, the examples have shown opening files asynchronously because this method is more popular for iPhone and mobile development. When you open files asynchronously, all file operations occur on a separate thread from the main one. This will allow your interface and application to remain responsive when performing file commands.

However, there are times when you want to pause the main thread while you are writing to your file. One example is exiting. If you try to write code asynchronously when you receive an exiting event from your application, your data may not be written before the application closes. In this example, there is no guarantee that your data will be written synchronously either because iOS may halt your code and exit in the case of a crash. However, you have a better chance of it occurring if your data is written synchronously.

To open files synchronously, simply call the `open()` method on the `FileStream` class instead of the `openAsync()` method. With the `open()` method, you can either read the file or write to the file immediately afterwards. There is no need to register for complete events. Your application will pause until all the data in the file is read into the buffer before continuing to execute the next line of code in your application.

If you are reading or writing very small pieces of data, this method may work well enough for you, as you may not see a significant delay in responsiveness in your application. It also makes your development easier as you do not have to worry about adding and removing event listeners for when the data is available.

Handle Files Synchronously

1. Create a `File` variable, such as `var file:File`.
2. Select a directory for your file, such as `File.documentsDirectory`.
3. Resolve the path, such as `resolvePath();`.
4. Create a name for your file, such as `preferences.txt`.

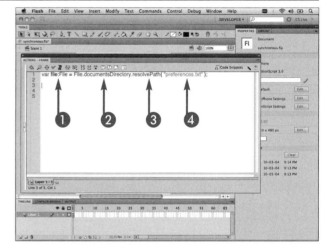

5. Create a `FileStream` variable, such as `var stream:FileStream = new FileStream();`.
6. Create a `try` block.
7. Create a `catch` block.
8. Catch for any errors, such as `error:Error`.

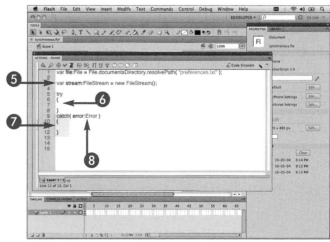

⑨ Open the stream, such as `stream.open();`.

⑩ Select the file to open.

⑪ Open the file in the `FileMode.READ` mode.

⑫ Create a `String` variable, such as `var str:String`.

⑬ Read the entire file, such as `stream.readUTFBytes( stream.bytesAvailable );`.

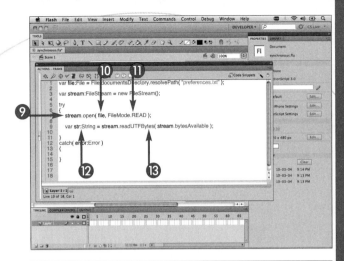

⑭ Output the file, such as `trace( str );`.

⑮ Close the stream, such as `stream.close();`.

⑯ Output any errors that are caught, such as `trace( error.message );`.

The file is opened and read synchronously.

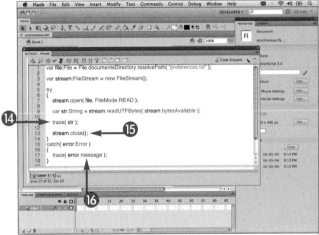

Extra

When creating a `File` object for a specific file location, you must make sure that the entire path to the file is valid. For example, the following `File` variable would throw an error because the full path to the file does not exists:

```
var file:File = File.userDirectory.resolvePath( "AppData/data.xml" );
```

The reason this would fail is because the AppData directory does not exist in the user directory of the application.

To make sure that your file is successfully created in this location, you would need to check to see if the directory existed and if it did not, create it. You can use the following syntax to create a directory:

```
var file:File = File.userDirectory.resolvePath( "AppData" );
file.createDirectory();
```

Copy Files

As mentioned earlier in this chapter, you cannot write files to the same directory as your application. The reason for this is that each application needs to be digitally signed when it is compiled. Adding or removing files from this directory may cause the application to be no longer valid and could cause it to no longer launch.

There may be times that you want to bundle files with your application and have them be updated throughout the life of the application. A good example of this would be when you are using a SQLite database. You may want to bundle a database with your application that contains information for your application. If you kept the database in the application directory, you would never be able to update or add new data to it.

To work around this issue, you can copy your file from the application directory to a directory to which you have write access. It is important that you copy and not move the file as this may cause the application to be invalid.

There are two methods that you can use to copy a file to a new location: The `copyTo()` and `copyToAsync()` methods of the `File` class give you the ability to copy your files synchronously or asynchronously, respectively. Both of these methods also allow you to specify whether to overwrite the file if it already exists.

When your application launches, it can check to see if the file already exists in a write access directory, and if it does not, it can copy the file from your application directory.

Copy Files

Note: *Do not forget to bundle the source file with your application. For more details, see the section "Bundle Images with Your Application" in Chapter 6.*

Note: *Bundling a text file uses the same process as bundling an image.*

1. Create a `File` variable, such as `var sourcefile:File`.
2. Select a directory for your file, such as `File.applicationsDirectory.resolvePath();`.
3. Create a name for your file, such as `preferences.txt`.

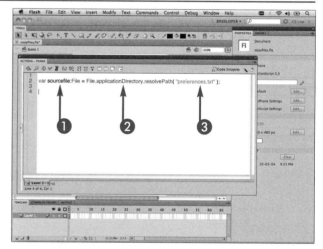

4. Create a `File` variable, such as `var destfile:File`.
5. Select a directory for your file, such as `File.documentsDirectory.resolvePath();`.
6. Create a name for your file, such as `preferences.txt`.

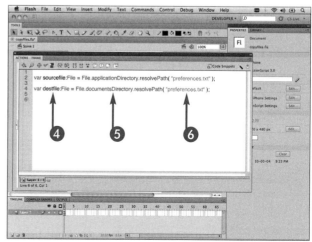

7 Register a listener for the complete event, such as `sourcefile.addEventListener( Event.COMPLETE, copyComplete );`.

8 Create an event handler method, such as `copyComplete`.

9 Output that the file has been successfully copied, such as `trace( "copy complete" );`.

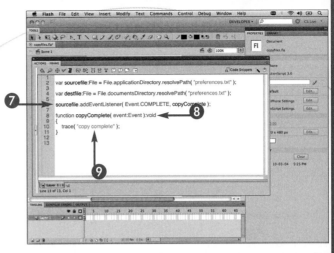

10 Copy the source file, such as `sourcefile.copyToAsync();`.

11 Select the destination file to be copied.

12 Overwrite the file if it already exists.

The file is copied.

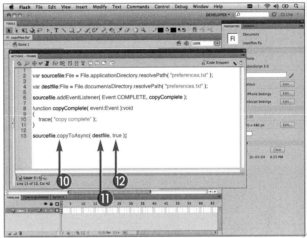

Apply It

It is possible that you will be copying files that are big in size. Because copying files from your application bundle will most likely occur during the startup procedure of your application, you will most likely favor the `copyToAysnc()` method. This will allow your application to start up as quickly as possible. However, if your application requires those files to be in the new location before your application can be interacted with, you will want to provide the user visual feedback that a copy is occurring. You can register a listener for the complete event on the `File` instance in order to know when the copy has completed successfully:

```
sourcefile.copyToAsync( destinationFile, true );
sourcefile.addEventListener( Event.COMPLETE, copyComplete );
function copyComplete( event:Event ):void
{
trace( "copy complete" );
}
```

Retrieve Your Current Location

The iPhone 3G and the iPhone 3GS both have the ability to retrieve your location via the onboard GPS chip on the device. The chip on the iPhone 3G is designed to find out where you currently are, and its accuracy can be off. The 3GS model of the iPhone does have a better GPS chip as well as a digital compass on board. Both devices also use cell tower triangulation in order to improve the accuracy of your location. The iPod touch does not come with a GPS chip onboard; however, when connected to a WiFi hotspot, its connection can be used to determine your location.

The Geolocation class allows you to retrieve information about your current location from your device. The geographical location is identified by longitude and latitude positions, which are returned in the WGS 84 standard format.

The Geolocation class dispatches an update event when a change in geographical location is detected. Each update dispatches a new GeolocationEvent, which returns properties of your current location, such as the longitude, latitude, altitude, speed, and timestamp.

You can set the time interval for updates in milliseconds by calling the Geolocation.setRequestedUpdate Interval() method. The update interval is used only as a hint to conserve the battery power, and the actual time between updates may be greater or lesser than this value. Omitting this call will result in the device using a default interval for its updates.

To check to see if the device and platform your application is running on support retrieving the current geographical location, you can see if the Geolocation.isSupported property is true.

Retrieve Your Current Location

1. Create a Geolocation instance, such as var geo:Geolocation;.

2. Check if Geolocation is supported, such as if(Geolocation.isSupported).

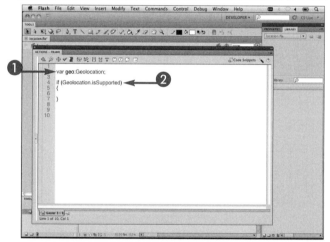

3. Instantiate a new Geolocation variable, such as geo = new Geolocation();.

4. Set the update interval, such as geo.setRequestedUpdate Interval(100);.

5. Register an event listener for the update event, such as geo.addEvent Listener(GeolocationEvent. UPDATE, geolocationUpdate Handler);.

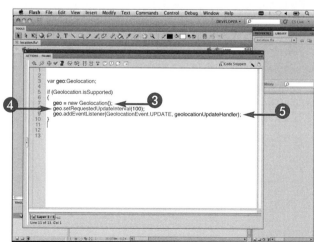

6 Create an `else` statement.

7 Output that `Geolocation` is not supported, such as `trace( "No geolocation support." );`.

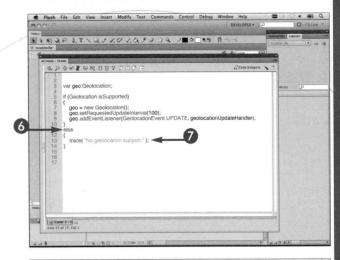

8 Create an event handler function for the update event, such as `geolocationUpdateHandler`.

9 Output the current latitude, such as `trace( "latitude:" + event.latitude.toString() + "°" );`.

10 Output the current longitude, such as `trace("longitude:" + event.longitude.toString() + "°");`.

The current latitude and longitude will be shown in the Output panel.

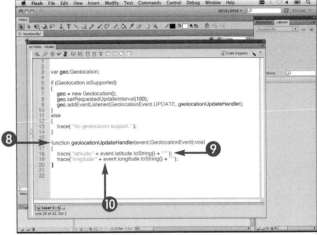

Extra

There is a great application in the App Store called *XSensor*, which is developed by Crossbow Technology, Inc. There is both a free version and a paid version of the application. With the paid version you get logging of your data. The XSensor application gives you all the data for all the sensors on your device. The Accels section of the application shows the information from the accelerometer on your device. The values of all three axes are visible as well as a graph that shows each axis.

The GPS section gives you all the information that the `GeolocationEvent` class would give you during an update from a `Geolocation` instance. You can easily see the latitude and longitude and the accuracy of your current geographical location. It also gives you the time of the last update as well as your speed in meters per second.

This information comes in really handy as it enables you to insert dummy data into your application, allowing you to test it on your computer before installing it on your device.

Map Your Location

Included in the iPhone SDK, there is a framework called MapKit. This framework gives you the ability to include and interact with maps in your application. This is very similar to having the Maps application embedded in your application. Currently, this framework is not included in the Flash CS5 iPhone Packager, but there are many mapping APIs available for Flash.

The Google Maps API is the most popular Flash mapping API; however, because of the way it is developed, it will not run on your device. Yahoo! has a map API for Flash that can be used in your iPhone applications. You will need to download the YahooMap.swc from the Yahoo! developer site at http://developer.yahoo.com/flash/maps/, as well as register your application in order to get an application ID. Yahoo! has lots of great samples that you can use to help get you started.

You can add the YahooMap.swc file in the Advanced ActionScript 3.0 Settings dialog box on the Library Path tab. It is important after you have included it with your project that you set it to be merged into the rest of your code when your project is compiled. Skipping this step will result in your application not running.

The Yahoo! map component has all the map features you would expect, such as zoom, pan, and the ability to switch between map, satellite, and hybrid style maps. After the map has been initialized, you can center it at a specified latitude and longitude position.

Make sure to read the license that comes with the API to make sure that you have the proper permission to include it in your application.

Map Your Location

Add the Map API to Your Application

Note: First, you must download and import the YahooMap.swc file from http://developer.yahoo.com/flash/maps/.

1. Click File → ActionScript Settings.
2. Click the Library Path tab.
3. Click the new Browse to SWC File button.
4. Click the SWC file that you imported.
5. Click here and select Merged into Code.
6. Click OK.

Display the Map at Your Current Location

7. In the Actions panel, import the necessary classes.
8. Create a `Geolocation` instance, such as `var geo:Geolocation = new Geolocation();`.
9. Create a new `YahooMap` instance, such as `var map:YahooMap = new YahooMap();`.
10. Register an initialize event listener, such as `map.addEventListener( YahooMapEvent.MAP_INITIALIZE, mapInitialized );`.
11. Initialize the map, such as `map.init();`.
12. Enter your Yahoo! App ID.
13. Enter the width and height of the map.

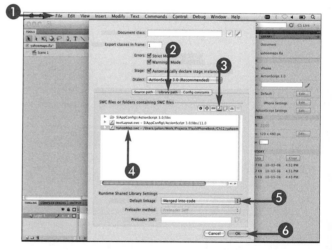

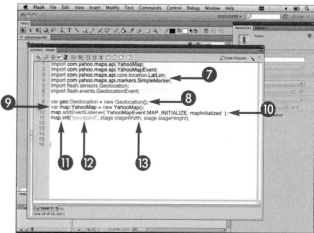

⑭ Create an event handler, such as `mapInitialize`.

⑮ Set the initial map zoom level, such as `map.zoomLevel = 4;`.

⑯ Register for `Geolocation` updates, such as `geo.addEventListener ( GeolocationEvent.UPDATE, onGeoUpdate );`.

⑰ Create an event handler, such as `onGeoUpdate`.

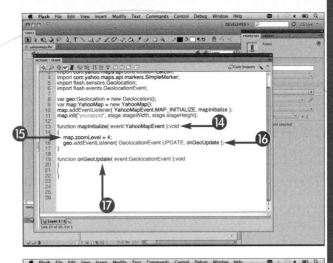

⑱ Create a new latitude/longitude variable, such as `var latlon:LatLon = new LatLon();`.

⑲ Enter the current latitude.

⑳ Enter the current longitude.

㉑ Center the map to the current position, such as `map.centerLatLon = latlon;`.

㉒ Add a pan control, such as `map.addPanControl();`.

㉓ Add a zoom control, such as `map.addZoomWidget();`.

㉔ Add the map to the Stage, such as `addChild(map);`.

The map will be displayed centered on the current location.

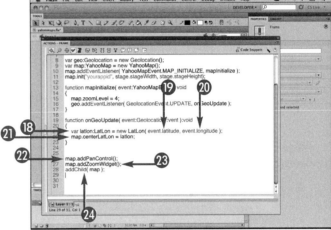

Apply It

There will be times when you want to find the latitude and longitude of a specific address. Calling the `geocode()` method on an `Address` instance will return a `LatLon` instance for that address. There is the possibility that more than one result will be returned to you. The following example automatically grabs the first one that is returned as this should be the most accurate:

```
var address:Address = new Address("1 Infinite Loop Cupertino, CA 95014");
address.addEventListener(GeocoderEvent.GEOCODER_SUCCESS, geocodeAddress);
address.geocode();
function geocodeAddress (event:GeocoderEvent):void
{
var address:Address = event.target as Address;
var result:GeocoderResult = address.geocoderResultSet.firstResult;
map.zoomLevel = result.zoomLevel;
map.centerLatLon = result.latlon;
}
```

continued

Map Your Location (continued)

The Yahoo! maps API also allows you to add a marker to the map at a given latitude and longitude location. Using the `Geolocation` class, you can retrieve the current geographical location of the device and place a marker at this location on the map.

Registering an event for the `Geolocation` class update event will allow you to update the position of your marker as your geographical location changes. The `MarkerManager` class in the Yahoo! API has a `resetPosition()` method, which enables you to update the position of a marker. This method is very useful as it saves you from having to remove the original marker and adding a new one at the new location.

The `SimpleMarker` class in the Yahoo! API is a very simple marker that you can add to the Stage. It is not the nicest-looking marker, though, so you will likely want to add your own custom graphic. To create your own custom marker, create a `MovieClip` symbol in the Library and center your graphic in it. Then export it for ActionScript and set the base class to `com.yahoo.maps.api.markers.Marker`. Your custom marker graphic will then be a subclass of the `Marker` class and have all the abilities that other map markers do.

As well as update the position of your marker when your geographical location changes, you will want to center the map to the same location. Keeping your marker centered will ensure that your marker is never off the screen unless you manually pan the map. To pan the map, place a finger on the map and drag it in the direction that you want to move.

Map Your Location (continued)

Add a Marker to Your Map

1 Import a marker image to the Library.

Note: *See Chapter 6, "Working with Images," for more information.*

2 Create a `MovieClip` symbol, such as `marker_mc`.

Note: *See Chapter 2, "Getting Started with Flash CS5," for more information.*

3 Center the image to the registration point of the movie clip.

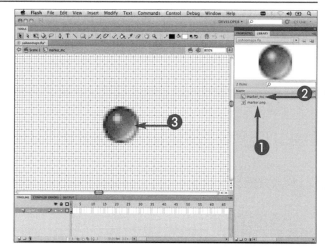

4 Right-click the `MovieClip` symbol and click Properties.

The Symbol Properties dialog box appears.

5 Click Export for ActionScript.

6 Give your symbol a class name, such as `CustomMarker`.

7 Give your symbol the base class of `com.yahoo.maps.api.markers.Marker`.

8 Click OK.

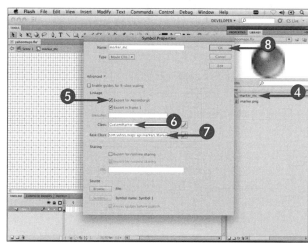

214

9 In the Actions panel, create an instance of your marker, such as `var marker:CustomMarker = new CustomMarker();`.

10 Set the `latlon` property of the marker, such as `marker.latlon = latlon;`.

11 Add the marker to the map, such as `map.markerManager.addMarker(marker);`.

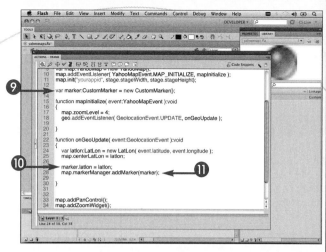

12 Publish and install your application on your device.

• Your marker will appear on the map, at the current geographical location.

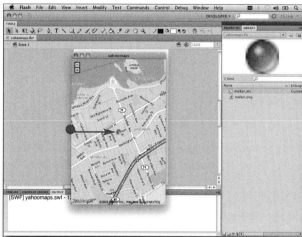

Apply It

The Yahoo! maps API also has the capability to perform a local search for specific items. For example, you could search for all Starbucks locations nearby. By creating an instance of the `LocalSearch` class, you can use the `searchLocal()` method to perform your search. An array of `LocalSearchItem` instances will be returned, which you can use in order to place a marker on the map.

```
var localSearch:LocalSearch = new LocalSearch();
localSearch.addEventListener( LocalSearchEvent.SEARCH_SUCCESS, searchSuccess );
localSearch.searchLocal( "Starbucks", map.zoomLevel, map.centerLatLon, 5 );
function searchSuccess( event:LocalSearchEvent ):void{
var localResults:LocalSearchResults = event.data as LocalSearchResults;
var results:Array = localResults.results;
}
```

Determine Your Speed

The Geolocation class does a great job of enabling you to retrieve your current geographic location from the device. When the device detects a change in location, a GeolocationEvent is dispatched, and an instance containing information about the update is passed to the event handler function. As well as determine the latitude and longitude of your current location from the GeolocationEvent instance, you can determine the speed at which you are traveling.

The GeolocationEvent.speed property returns a number that represents your speed in meters per second. The example in the steps below shows the equation for converting the number to kilometers per hour. Use the following equation if you want to convert it to miles per hour:

```
var mph:Number = Math.
 round((mps*360000)/160934.4));
```

The accuracy of the speed property will vary depending on the accuracy of the GPS data that you receive. It will also vary between the different devices. It has been my experience that it has been fairly accurate.

The speed property should be used more as a novelty instead of relying on it for very accurate data. For example, using it to see what your top speed is during a run or a bike ride is far more effective than using it as a replacement for the speedometer in your car.

You can also determine the time of when each update occurs. The timestamp property on the GeolocationEvent class returns the number of milliseconds since the runtime was initialized. For example, if your application captures its first update event 30 seconds after it started, the timestamp property would return 30000.

Determine Your Speed

1. Click the Text tool.
2. Draw a text field on the Stage.
3. Click here and select Classic Text.
4. Click here and select Dynamic Text.
5. Give the text field an instance name, such as speed_txt.

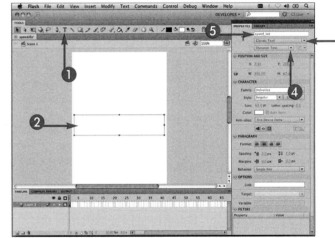

6. In the Actions panel, check to see if Geolocation is supported, such as if(Geolocation.isSupported){}.
7. Create an else statement if it is not supported, such as else{}.

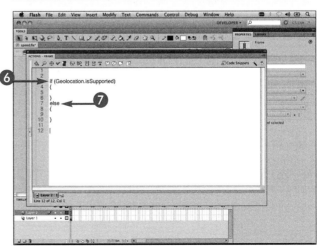

8 Create a `Geolocation` variable, such as `var geo:Geolocation = new Geolocation();`.

9 Add an event listener, such as `geo.addEventListener (GeolocationEvent.UPDATE, geolocationUpdateHandler);`.

10 Output a message if `Geolocation` is not supported, such as `trace( "No geolocation support" );`.

11 Create an event handler function.

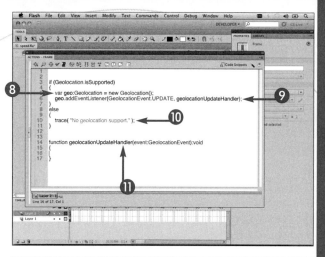

12 Get the current speed, such as `var mps:Number = event.speed;`.

13 Convert the speed to km/h, such as `var kph:Number = Math.round( ( mps * 3600 ) / 1000 );`.

14 Set the text of the text field to the current speed, such as `speed_txt.text = kph + " km/h";`.

The current speed of the device will be displayed in the text field.

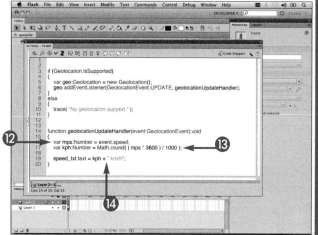

Apply It

The `GeolocationEvent` class also contains a property that represents your altitude in meters. Combining both `altitude` and `speed` properties, you could measure how fast you ascend or descend in altitude. For example, you could measure your top speed skiing down a hill while measuring the change in altitude from the top to the bottom of the run.

```
var geo:Geolocation = new Geolocation();
geo.addEventListener(GeolocationEvent.UPDATE, geolocationUpdateHandler);
function geolocationUpdateHandler(event:GeolocationEvent):void
{
    trace( "altitude", event. altitude );
}
```

Retrieve a List of Contacts

Included in Apple's iPhone SDK is the Address Book framework, which gives you the ability to access, create, and edit your contacts. It also contains the ability to present the user with similar user interface elements to that of the Contacts application. Giving users this familiarity makes them more confident when editing their contacts through your application. Unfortunately, currently the Flash CS5 iPhone Packager does not include the Address Book framework.

However, there is one workaround to interacting with the user's contacts. All the information for the Contacts application is stored in a SQLite database on the device. With the SQLite classes included in the iPhone Packager, you are able to connect to this database to read the information. However, the database is located outside of the application sandbox, and Apple may reject your application for accessing files outside of its sandbox.

The information for a contact is spread out over multiple tables in the database. The ABPerson table contains a list of all the contacts. This table provides columns for the contact's first name, last name, middle name, organization, department, and job title, among others. The ROWID column is used as a unique ID for that contact. You will use this ID to find other information about the contact in other tables in the database.

When retrieving the complete list of contacts, you will want to return only what is necessary to display them in a list. Contact lists on your device show only their first and last names to start, and they show detailed information only when a contact is selected. You will also need the ID for that user in order to retrieve the details at a later time.

Retrieve a List of Contacts

1. Create a new `File` instance to the database, such as `var file:File = new File("/private/var/mobile/Library/AddressBook/AddressBook.sqlitedb");`.

2. Create a `SQLConnection` instance, such as `var connection:SQLConnection = new SQLConnection();`.

3. Listen for the open event, such as `connection.addEventListener( SQLEvent.OPEN, onConnect );`.

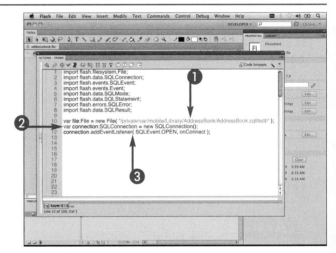

4. Open the connection, such as `connection.openAsync( file, SQLMode.READ );`.

5. Create an event handler for the open listener.

6. Create a new `SQLStatement` instance, such as `var statement:SQLStatement = new SQLStatement();`.

7. Set the connection, such as `statement.sqlConnection = connection;`.

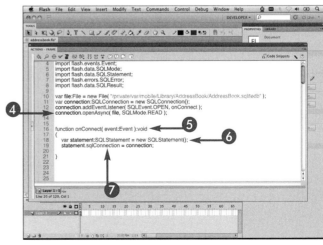

8. Select the data from the database, such as `statement.text = "SELECT ROWID, First, Name FROM ABPerson";`.

9. Add a listener for the result event, such as `statement.addEventListener ( SQLEvent.RESULT, contactListResult );`.

10. Execute the SQL statement, such as `statement.execute();`.

11. Create the event handler function.

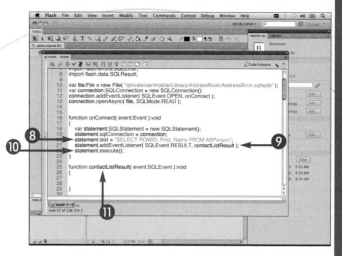

12. Get the results, such as `var result: SQLResult = SQLStatement (event.target).getResult();`.

13. Loop through the result data, such as `for( var i:int = 0; i<result.data.length; i++ ){}`.

14. Create a reference to the row data, such as `var item:Object = result.data[ i ];`.

15. Output the row data, such as `trace ( item.ROWID, item.First, item.Last );`.

The first and last name of every contact in your Address Book will be shown in the Output panel.

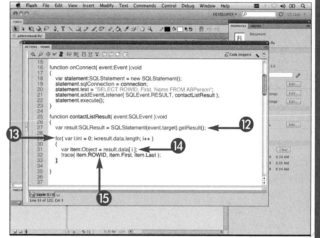

Extra

If you know the name of the contact you are looking for, you can get the data from checking the value of the First column and the value of the Last column. This will save you from having to select all the contacts in the database and look through them on the client. You can use the following syntax to perform a SELECT statement on specific contacts:

```
var sql:String = "SELECT ROWID FROM ABPerson WHERE First='firstname' AND Last='lastname'";
```

Retrieve a Contact's Details

The details for a contact are spread out over multiple tables in the database. This means that you will need to create multiple selects in order to gather all the details. To do so, you will need to retrieve the contact's unique ID in the ROWID column of the ABPerson table. You can select all the contacts from the table to find one contact. For more details on how to accomplish this, see the preceding section, "Retrieve a List of Contacts."

The ABMultiValue table in the database stores all the items, which can have multiple entries, such as phone numbers, email addresses, addresses, and Web sites. It also has a record_id column in the table, which represents the ROWID of the contact from the ABPerson table.

The ABMultiValueEntry table in the database contains all the information for the address and instant message accounts of the contact. There is also an entry in the ABMultiValue table for each of the contact's addresses and instant message accounts. That entry's UID column value is used to check against the parent_id column of the ABMultiValueEntry table in order to select all the information for the given item.

Combining all the contact's data from the ABPerson, ABMultiValue, and ABMultiValueEntry tables will give you all the detailed information for that contact.

There are ways to select all the data in one SELECT statement using the JOIN keyword; however, this can and would get very complicated. It is much easier to do a single SELECT statement for each table based on the previous results.

Retrieve a Contact's Details

Note: See the preceding section, "Retrieve a List of Contacts," to get a list of available contacts in order to get their details.

1. Create a function to get the details of a contact by ID, such as `function getContactDetails(id:int): void {}`.

2. Create a SQLStatement variable.

3. Set the SQL connection.

4. Select the details based on ID, such as `statement.text = "SELECT * FROM ABMultiValue WHERE record_id=" + id;`.

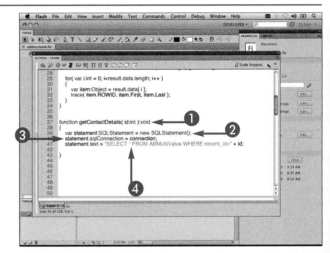

5. Add a listener for the results event, such as `statement.addEventListener ( SQLEvent.RESULT, contactDetailsResult );`.

6. Execute the statement, such as `statement.execute();`.

7. Create an event handler function.

8. Get the SQL results, such as `var result:SQLResult = SQLStatement(event.target). getResult();`.

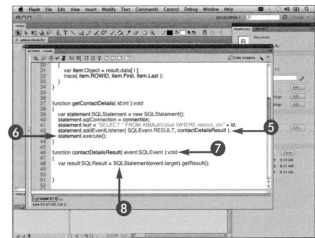

⑨ Loop through the results.

⑩ Create an `Object` variable for the result, such as `var item:Object = result.data[ i ];`.

⑪ Create a `switch` statement on the `item.property` property.

⑫ Create a `case` for `3` to check for a phone number, such as `case 3:`.

⑬ Output the phone number, such as `trace( "number", item.value);`.

⑭ Add a break statement, such as `break;`.

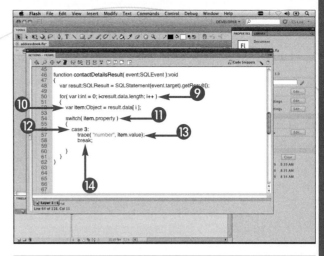

⑮ Create a `case` for `4` to check for an email address, such as `case 4:`.

⑯ Output the email address, such as `trace( "email", item.value);`.

⑰ Add a break statement, such as `break;`.

⑱ Create a `case` for `22` to check for a Web site URL, such as `case 22:`.

⑲ Output the Web site, such as `trace ( "website", item.value);`.

⑳ Add a break statement, such as `break;`.

The contact's details will be shown in the Output panel.

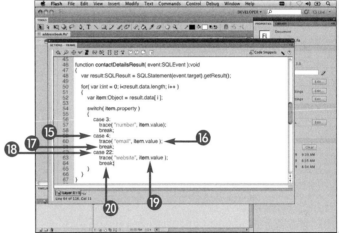

Apply It

There is also an `IPhoneAddressBook` class as part of the as3iphone open source project, at http://code.google.com/p/as3iphone/. This class does a lot of what was shown in the example and more. It makes retrieving contacts much easier. Below is an example of getting a list of contacts:

```
var ab:IPhoneAddressBook = new IPhoneAddressBook();
ab.addEventListener( Event.OPEN, onOpen );
ab.open();
function onOpen( event:Event ):void{
ab.addEventListener( AddressBookEvent.CONTACT_LIST, onContactList );
ab.getContactList();
}
function onContactList(event:AddressBookEvent):void{
var data:Array = event.data;
}
```

Retrieve Phone Number Favorites

Many of the preferences for your device are stored in property list files on it. A *property list file*, a file with the .plist extension, comes in two forms. The first form is a standard XML text-based file. It can be read and converted to an XML object, just as you normally would for any other .xml file. The second form is a binary version of the text-based version of the file. The binary version allows the file to be compressed in order to conserve space on the local file system. The binary version of the file is more difficult to read, as you must know the file format in order to decode all the properties in it.

One of the preferences that are stored in a property list file is your Favorites in the Phone application, which is available on only an iPhone and not an iPod touch. The file has an array of dictionary objects, each representing an entry in your Favorites list. Each entry contains the name of the contact, the phone number, and the unique ID in the ABPerson table in the Address Book SQLite database table.

The property list file that stores the Favorites is a binary version of the file. Understanding the file format is a complicated task; however, included with the source code of the book there is a BPlistParser class. This class will take care of reading and parsing a specified file. You can also download the latest version of the class from the Google code project at http://code.google.com/p/as3iphone/.

Retrieve Phone Number Favorites

① In a Web browser, go to http://code.google.com/p/as3iphone/source/checkout.

② Follow the instructions for checking out the latest files with svn.

③ In Flash, open the Actions panel.

④ Import the File class, such as import flash.filesystem.File;.

⑤ Create a new File variable, such as var file:File = new File();.

⑥ Enter the path to the Favorites file, such as "/private/var/mobile/Library/Preferences/com.apple.locationd.plist".

⑦ Create a new BPlistParser instance, such as var bplist:BPlistParser = new BPlistParser();.

⑧ Open the file, such as bplist.open(file);.

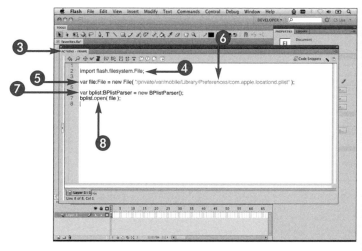

222

⑨ Get the root Array, such as var entries:Array = bplist.getRootArray();.

⑩ Iterate over all the items in the array.

⑪ Create a reference to the current entry, such as var entry:Dictionary = entries[i] as Dictionary;.

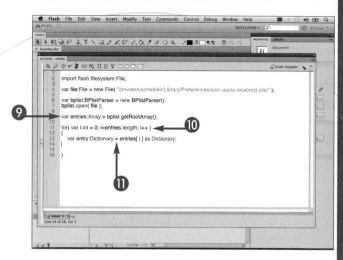

⑫ Create a String variable for the name, such as var name:String = entry["Name"] as String;.

⑬ Create a String variable for the number, such as var number:String = entry["Value"] as String;.

⑭ Create an int variable for the contact ID, such as var id:int = entry["ABUid"] as int;.

⑮ Output the name and number, such as trace(name, number);.

The name and number for each favorite will be shown in the Output panel.

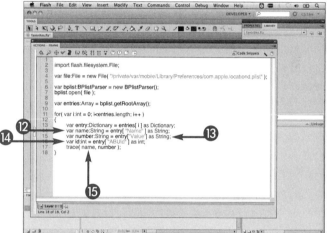

Extra

To get the full contact details of a contact in the Favorites list, you can use the ABUid property of the entry. This ID represents the value in the ROWID column in the ABPerson table of the Address Book SQLite database. For more information on how to connect to the Address Book database, see the section "Retrieve a List of Contacts" earlier in this chapter. The following is a sample SELECT statement that will enable you to select all the data for a contact based on his or her ABUid value:

```
var sql:String = "SELECT * FROM ABPerson WHERE ROWID=" + entry[ "ABUid"];
```

Check for an Internet Connection

Chances are your application will at some point require a connection to the Internet. It may be to simply send analytic tracking for user interactions or to submit a score to an online leaderboard. It could also be as complicated as a multiplayer game over a socket connection or streaming video from a media server, such as Flash Media Server. It is a good practice to check for the presence of an Internet connection before attempting to send data.

If an Internet connection is not found, you can hide the portions of your application that require it — for example, not displaying a Submit button in order to submit a score. You can also save the score locally for the user, and next time that he or she launches the application with an Internet connection, you can submit the scores.

The `URLMonitor` and `SocketMonitor` classes enable you to monitor the availability of a specific host or service. The `URLMonitor` class is used to monitor the availability of a specified `URLRequest` instance. Calling the `start()` method of the `URLMonitor` class will attempt to hit the specified URL. After the monitor has determined whether the URL is available or not, a `StatusEvent.STATUS` event will be fired. The value of the `code` property for the event instance will be set to `"Service.available"` if the URL is available and `"Service.unavailable"` if it is unavailable. However, it is best practice to check the value of the `available` property of your `URLMonitor` instance.

Setting the `pollInterval` property of a `URLMonitor` instance will cause it to check the availability of the URL at the specified interval in milliseconds. Calling the `stop()` method will stop the monitoring of the service.

Check for an Internet Connection

1. Click File → ActionScript Settings.

   The ActionScript Settings dialog box appears.

2. Click the Library Path tab.

3. Click the new Browse to SWC File button.

4. Click the aircore.swc file in the Adobe Flash CS5/AIK2.0/frameworks/libs/air folder.

5. Click here and select Merged into Code.

6. Click OK.

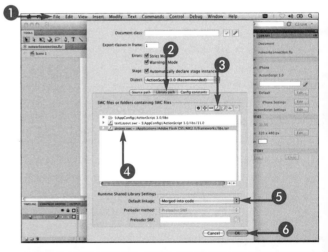

7. In the Actions panel, create a new `URLRequest` variable, such as `var request:URLRequest = new URLRequest();`.

8. Enter a valid URL to monitor, such as `http://www.google.com`.

9. Create a new `URLMonitor` variable, such as `var monitor:URLMonitor = new URLMonitor();`.

10. Enter the `request` variable in the constructor.

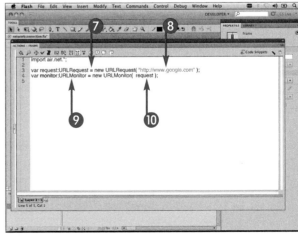

⓫ Add a listener to the monitor, such as `monitor.addEventListener();`.

⓬ Listen for the `StatusEvent.STATUS` event.

⓭ Enter an event handler method, such as `onStatus`.

⓮ Start the monitor, such as `monitor.start();`.

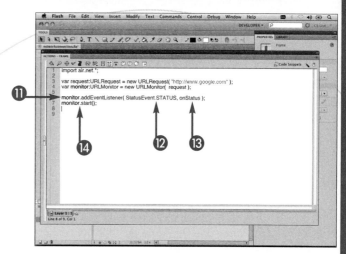

⓯ Create an event handler function.

⓰ Stop the monitor, such as `monitor.stop();`.

⓱ Output the monitor's availability, such as `trace( monitor.available);`.

The availability of the service will be shown in the Output panel.

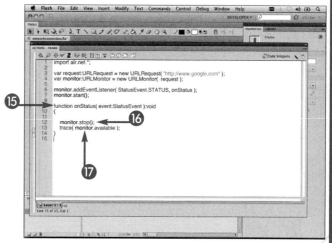

Apply It

The `SocketMonitor` class works the same way the `URLMonitor` class works, except instead of monitoring a URL or HTTP service, it monitors a host and port. Both classes are subclasses of the `ServiceMonitor` class and share many of the same properties and methods. When creating a `SocketServer` instance, you specify the host and the port in its constructor:

```
var monitor:SocketMonitor = new SocketMonitor("192.168.1.102", 4444);
monitor.addEventListener( StatusEvent.STATUS, onStatus );
monitor.start();
function onStatus( event:StatusEvent ):void
{
trace( monitor.available );
}
```

Check for a Persistent WiFi Connection

There are some applications that require a constant Internet connection. For example, a multiplayer game requires an Internet connection to connect to other users. Another example is a podcast or video application that only streams the data over the air and does not cache it on the device. If you develop an application that does require a constant Internet connection, you will want to check if the user has his or her device in Airplane mode. *Airplane mode* turns off all sending and receiving of data so that you can still use your device on an airplane.

The UIRequiresPersistentWiFi key can be added to the Info.plist file of your application to alert the user that an Internet connection is required. The Info.plist file is an XML-based property list file that is bundled with your application. It contains configuration details about your application, such as the application version number, the default orientation, and whether to show or hide the status bar. The Info.plist list file generated based on the information in your application descriptor XML file and details from the iPhone Packager.

The UIRequiresPersistentWiFi key is a Boolean value that, when true, will alert the user to turn off Airplane mode. You can add additional items to your application descriptor XML file that will be added to the Info.plist file when it is generated upon compilation of your application.

All additional Info.plist keys are added to the <InfoAdditions> node of the <iPhone> node. These nodes are not created by default by Flash, and you will have to add them yourself. All the additional keys are also placed in CDATA in order to be ignored by the application descriptor parser.

Check for a Persistent WiFi Connection

Check for a Persistent WiFi Connection

1. Open the application descriptor file for your application.

   Note: You can find this file in the same directory as the SWF for your application.

2. Create an `<iPhone></iPhone>` XML node.

3. Create an `<InfoAdditions></InfoAdditions>` XML node.

4. Create a CDATA tag, such as `<![CDATA[ ]]>`.

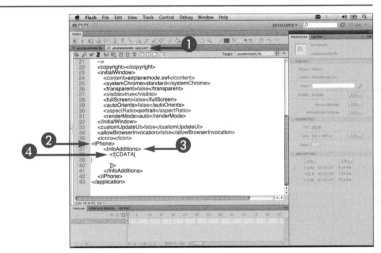

5. Add the UIRequiresPersistentWiFi key, such as `<key>UIRequiresPersistentWiFi</key>`.

6. Set the key to `<true/>`.

   The UIRequiresPersistentWiFi key will be added to the Info.plist file when the application is compiled.

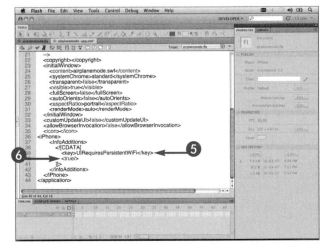

Open the Info.plist File

7 In a Finder window or Explorer, rename your .ipa file to a .zip file.

8 Double-click the .zip file to extract it.

9 Click here to expand the Payload folder.

10 Right-click the .app file.

11 Click Show Package Contents.

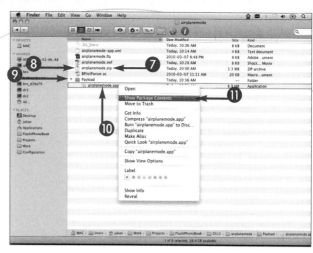

Note: Inside your application bundle, you will find the Info.plist file.

12 Open the Info.plist file.

● The `UIRequiresPersistentWiFi` key is set to true.

Apply It

There is also another way to check if the device is in Airplane mode. There is a preference binary property list file that is located on the device that gets updated when the device enters and exits Airplane mode. You have to check this file only when your application starts up, as it is not possible to change Airplane mode while an application is running. You can use the `BPlistParser` class that you used to retrieve the phone number favorites:

```
var file:File = new File( "/private/var/mobile/Library/Preferences/com.apple.BTServer.
  airplane.plist" );
var bplist:BPlistParser = new BPlistParser();
bplist.open( file );
var dict:Dictionary = bplist.getRootDictionary();
var airplanemode:String  = dict[ "airplaneMode" ];
if( airplanemode == "off" ){
trace( "airplane mode is off" );
}else if( airplanemode == "on" ){
trace( "airplane mode is on" );
}
```

Set the System Idle Mode

The greatest thing about the iPhone and iPod touch is having your applications and information with you at all times. Mobile devices are becoming more and more like minicomputers, and we depend on them just as much — if not more. The biggest obstacle for any portable device, laptops included, is battery life. The more your application does, the more it will drain the battery.

In order to help prolong battery life on the device, you can set it to go into Sleep mode if it does not register a touch on the screen. When the device goes into Sleep mode, the screen turns off, and the device locks. Because of the different ways that these devices allow users to interact, this can be an issue. For example, a game that uses only the accelerometer as its input will not register a screen touch, so the screen could turn off right in the middle of a game.

This scenario is not ideal, and you will need to tell the device to remain awake. When disabling the Sleep mode of the device, be sure to disable it only when necessary. For example, if you are building a game, only disable it during game play and when the game launches. This allows the device to go into Sleep mode when it is not critical that it stay awake. This will help conserve the battery power of your user's device.

The `NativeApplication.systemIdleMode` property allows you to tell the device to stay awake, as well as set it back to the normal mode of the device.

Set the System Idle Mode

Create Buttons

1. Create a `Button` symbol and add it to the Stage.

Note: See Chapter 2 for more information.

2. Give it an instance name, such as `awake_btn`.

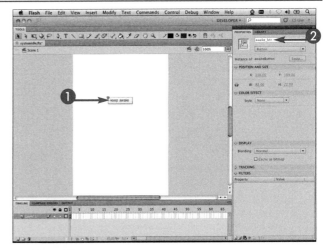

3. Create another `Button` symbol and add it to the Stage.

4. Give it an instance name, such as `normal_btn`.

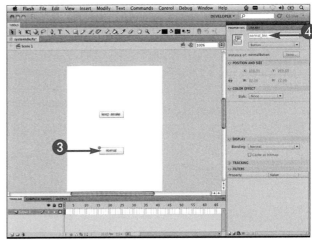

Have the Buttons Set Awake and Normal Idle Modes

5 In the Actions panel, add a listener for the click event, such as `normal_btn.addEventListener( MouseEvent.CLICK, onNormal );`.

6 Create an event handler function, such as `onNormal`.

7 Set the system Idle mode to normal, such as `NativeApplication.nativeApplication.systemIdleMode = SystemIdleMode.NORMAL;`.

8 Add a listener for the click event, such as `awake_btn.addEventListener( MouseEvent.CLICK, onAwake);`.

9 Create an event handler function, such as `onAwake`.

10 Set the system Idle mode to awake, such as `NativeApplication.nativeApplication.systemIdleMode = SystemIdleMode.KEEP_AWAKE;`.

When the `awake_btn` button is clicked, the device will remain awake.

When the `normal_btn` button is clicked, the device will sleep after a certain period of inactivity.

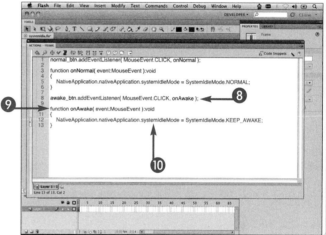

Apply It

When your application exits, the system Idle mode is reset back to normal, allowing the device to enter Sleep mode when needed. However, you should not rely on this at all times; it is best practice to return the system idle back to normal when your application exits. This will make sure that the device will return to Sleep mode as expected. You can listen for the exiting event of the application to determine when the application is about to close:

```
var application:NativeApplication = NativeApplication.nativeApplication;
application.addEventListener( Event.EXITING, onExit );
function onExit( e:Event ):void
{
application.removeEventListener( Event.EXITING, onExit );
application.systemIdleMode = SystemIdleMode.NORMAL;
}
```

Make Phone Calls

The iPhone has a custom URL scheme to enable developers to present users with a link to a telephone number. Creating a link to a telephone number is very similar to creating a link to a Web page. However, you substitute the http:// protocol with tel:. The tel URL scheme takes any valid phone number as its parameter. When navigating to a valid tel URL, the Phone application will launch, and the number will be dialed without prompting the user.

It is a good practice to show an alert to the users, allowing them to confirm that they want to dial the specified number. The Mail application does this when a telephone number is detected in an email.

The Phone application supports most, but not all, of the special characters in the tel URL scheme. For example, if a tel URL contains the * or # characters, the Phone application does not attempt to dial the corresponding phone number. This helps in preventing users from redirecting phone calls or changing the behavior of the Phone application. Also, if any space characters are detected in the URL string, the Phone application will not launch, as these spaces are not supported.

When dealing with an unknown source, such as allowing the user to enter a telephone number into an input text field, you should make sure that all special characters that are not valid characters in the URL are escaped properly. You can use the encodeURI() method to escape the URL string properly. When a string is encoded, all characters are encoded as UTF-8–escaped sequences unless they belong to a small subset of characters. To see the full list of characters, refer to the help files.

Make Phone Calls

① Create a Button symbol and place it on the Stage.

Note: See Chapter 2, "Getting Started with Flash CS5," for more information.

② Give it an instance name, such as phone_btn.

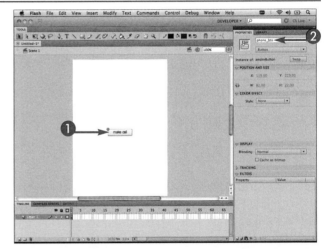

③ In the Actions panel, register a listener for the button click, such as phone_btn.addEventListener (MouseEvent.CLICK, onPhoneClick);.

④ Create an event handler function, such as onPhoneClick.

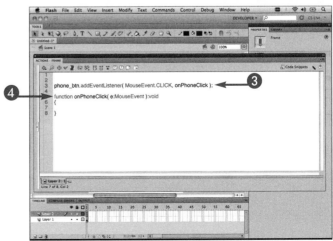

5. Create a String variable for the phone number, such as var number:String = "+1-555-555-5555";.

6. Create a String variable for the URL, such as var url:String = "tel:";.

7. Add the number variable to the url variable, such as + number.

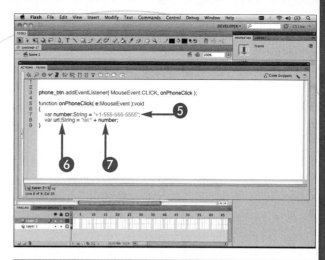

8. Create a URLRequest variable, such as var request:URLRequest = new URLRequest();.

9. Add the URL to the request.

10. Type **navigateToURL();**.

11. Add the request variable as an argument.

After you compile and install the application on a device, when the button is pressed, the Phone application will launch.

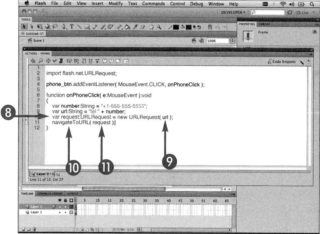

Extra

You can also place the telephone number as a link in a text field that renders HTML text. To set the text field to render HTML text, you must select the Render as HTML button in the Properties panel of the text field. It is the middle button located directly underneath the Anti-Alias drop-down list. After the text field has been set to render HTML text, you can set the htmlText property of the text field to a String containing HTML text. Here is the syntax to create a telephone number link in a text field:

```
mytext_txt.htmlText = 'My telephone number is <u><a href="tel:+1-555-555-5555">1-555-555-5555
    </a></u>';
```

Make sure that the font for the text field is big enough to allow the user to tap the link. If the font is too small, users will have a hard time registering a tap on the URL portion of the text.

Open the Mail Application

Opening the Mail application to send a message is the same process as if you were in a browser. The `mailto` URL scheme allows you to compose a new email message with certain fields prefilled. In its simplest form, a `mailto` URL contains only a recipient's email address, which follows the `mailto:` protocol. Any additional fields that you want to have prefilled are added as query string parameters after the recipient's email address.

You can use the `to` query string parameter in order to add additional email addresses to the To: field of the email. The `cc` query string parameter allows you to add email addresses to the carbon copy field. The `bcc` query string parameter allows you to add email addresses to the blind carbon copy field. If you want to add multiple email addresses to any of these fields, you can separate additional email addresses by a comma.

You may have multiple email accounts associated with your iOS device and may want to select with which email address to send the email. However, iOS ignores the `mailto` URL scheme's `from` property, and the Mail application will use your default email address.

Setting the `subject` query string parameter fills the subject field of the email. The `body` query string parameter allows you to prefill the body of the email with text. Because the =, ?, and & characters are reserved in the `mailto` scheme, any parameters that contain these characters must be encoded. You will need to encode each parameter separately and not the entire URL as that will cause reserved characters that are used to separate parameters to be encoded as well.

Open the Mail Application

① Create a `Button` symbol and place it on the Stage.

Note: See Chapter 2 for more information.

② Give it an instance name, such as `email_btn`.

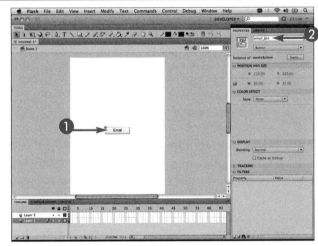

③ Register a listener for the button click, such as `email_btn.addEventListener( MouseEvent.CLICK, onEmailClick );`.

④ Create an event handler function, such as `onEmailClick`.

⑤ Create a `String` variable for the subject of the email, such as `var subject:String = "Try to beat my score";`.

⑥ Encode the variable, such as `encodeURI()`;.

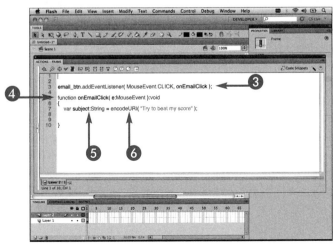

⑦ Create a String variable for the body of the email, such as var body:String = "I just scored 15429 in Sushi Toss! Can you beat my score?".

⑧ Encode the variable, such as encodeURI();.

⑨ Create a String variable, such as var url:String = "";.

⑩ Add the mailto: protocol.

⑪ Add an email address, such as name@email.com.

⑫ Add a ?.

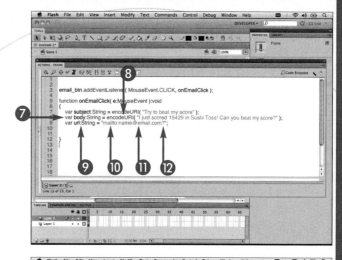

⑬ Add the subject parameter, such as subject=.

⑭ Add the subject variable.

⑮ Add the body parameter, such as &body=.

⑯ Add the body variable.

⑰ Create a URLRequest variable, such as var request:URLRequest = new URLRequest();.

⑱ Add the URL to the request.

⑲ Type **navigateToURL();**.

⑳ Add the request variable as an argument.

When the button is pressed on a device, the Mail application will launch with the specified data prefilled.

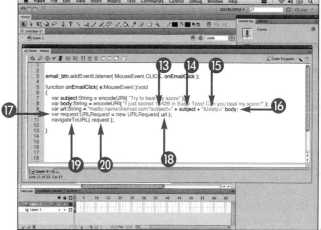

Apply It

Currently, the mailto URL scheme does not support attachments. A common integration point of the Mail application in an iPhone application is being able to send images with your email. This is seen in the Photos application that is preinstalled with your iOS device. However, this currently cannot be accomplished with the Flash CS5 iPhone Packager. You can, although, include HTML in the body of your email. The following is an example of HTML in the body of your email:

```
var html:String = "<a href='http://www.deleteaso.com'>My Blog</a>";
var subject:String = "link to my blog";
var url:String = "mailto:name@email.com?subject=" + encodeURI(subject) + "&body=" +
  encodeURI(html);
navigateToURL( new URLRequest( url ) );
```

233

Open the Maps Application

Chapter 12, "Using the Location, Contacts, and WiFi Features," discusses how you can integrate Yahoo! Maps inside of your application. Another alternative to this is being able to launch the native Maps application to a specific geographical location. The downfall to this is having your application exit, in order to show the map. Depending on the type of application you are creating, this may be acceptable. Unlike the other URL schemes in this chapter, the map URL does not start with a custom scheme identifier, such as `maps`. Instead, a regular `http` URL is used, targeted at the Google Maps servers. If the Maps application is not found, a browser will be launched and will go to the Google Maps page.

Most of the Google Map parameters are supported; however, using an unsupported parameter may cause the Maps application to fail to launch. The following are the available Google Map parameters: The `q` parameter is treated as if a query had been typed into the query box on the maps.google.com page. Note that `q=*` is not supported. The `near` parameter is the location part of the query. The `ll` parameter represents the latitude and longitude points in decimal format and separated by commas for the center of the map. The `t` parameter sets the type of map to display. Valid options for this parameter are `m` for maps, `k` for satellite, and `h` for hybrid. The `z` parameter sets the zoom level of the map ranging from 1–20. The `sll` parameter is the latitude and longitude points from which a business search should be performed. The `spn` parameter is the approximate latitude and longitude span.

Two parameters that cannot be included are `view=text` and `dirflg=r`.

Open the Maps Application

① Create a `Button` symbol and place it on the Stage.

Note: *See Chapter 2 for more information.*

② Give it an instance name, such as `map_btn`.

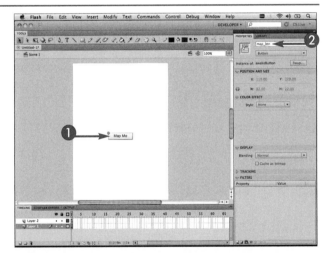

③ Register a listener for the button click, such as `map_btn.addEventListener ( MouseEvent.CLICK, onMapClick );`.

④ Create an event handler function, such as `onMapClick`.

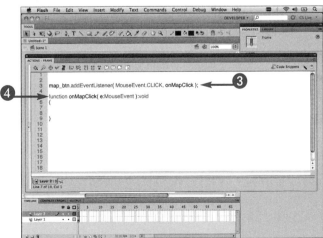

5. Create a String variable, such as `var url:String = "";`.

6. Add the Google Maps URL, such as `http://maps.google.com/maps`.

7. Add the latitude and longitude parameter, such as `?ll=`.

8. Enter a latitude, such as `40.770401835894084`.

9. Add a comma.

10. Enter a longitude, such as `-73.97420883178711`.

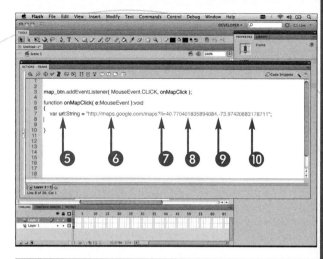

11. Create a URLRequest variable, such as `var request:URLRequest = new URLRequest();`.

12. Add the URL to the request.

13. Type **navigateToURL();**.

14. Add the `request` variable as an argument.

After you compile and install the application on a device, when the button is pressed, the Maps application will launch to the specified location.

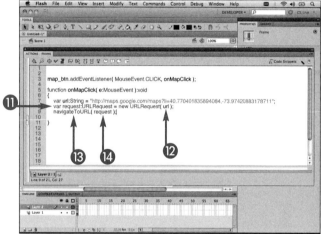

Apply It

As well as show the current location on the map, you can generate driving directions between two addresses. By setting the `saddr` parameter as the source address and the `daddr` parameter as the destination address, you can have driving directions displayed on the map between the two addresses. Here is an example of getting the driving directions between the Apple head office and the Adobe head office:

```
var saddr:String = encodeURI("1 Infinite Loop, Cupertino, CA, 95014");
var daddr:String = encodeURI("345 Park Ave, San Jose, CA 95110");
var url:String = "http://maps.google.com/maps?saddr="+saddr+"&daddr="+daddr;
navigateToURL( new URLRequest( url ) );
```

Open the Messages Application

The sms URL scheme makes it easy to open the Messages application from your applications. The Messages application is where you can send text messages or SMS messages to other SMS-enabled devices. The sms URL scheme can be used to simply open the Messages application, or you can open it to a new message with a phone number already prefilled.

The URL scheme starts with sms: and is followed by the phone number to compose the text message to. The phone number can contain digits 0 through 9 and the plus (+), hyphen (-), and period (.) characters. Any additional characters or text must not be included in the URL. If the sms URL does not have a phone number included with it, the Messages application will open to its default screen.

If the Messages application is not installed on the device, the appropriate error message will be shown to the user.

There are currently some options for the sms URL scheme that are not supported on the iPhone. For example, you can specify only one phone number per URL. In the specification, it states that you can add multiple phone numbers by separating them with commas. In reality, however, if a comma is used to separate two phone numbers, the entire string will be treated as one phone number.

Similarly to the mailto URL scheme, the sms URL scheme has a body parameter. This parameter is used to prefill the message portion of the text message. Unfortunately, this parameter is not supported by the iPhone Messages application, and if you use it, you may get unexpected results. To help get this feature added, I encourage you to go and file a Radar, Apple's feature request, on the iPhone Developer page.

Open the Messages Application

1. Create a Button symbol and place it on the Stage.

Note: *See Chapter 2 for more information.*

2. Give it an instance name, such as sms_btn.

3. Register a listener for the button click, such as sms_btn.addEventListener(MouseEvent.CLICK, onSMSClick);.

4. Create an event handler function, such as onSMSClick.

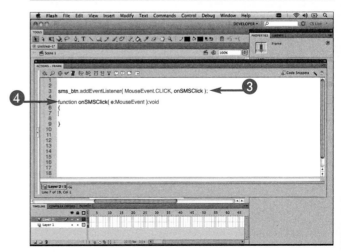

5. Create a String variable, such as `var url:String = "";`.
6. Add the sms URL protocol, such as `sms:`.
7. Add a phone number to text, such as 1-555-555-5555.

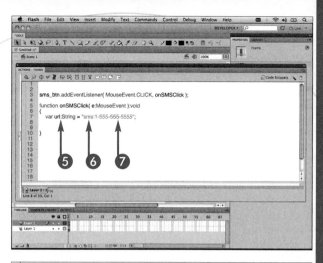

8. Create a URLRequest variable, such as `var request:URLRequest = new URLRequest();`.
9. Add the URL to the request.
10. Type **navigateToURL();**.
11. Add the request variable as an argument.

After you compile and install the application on a device, when the button is pressed, the Messages application will launch.

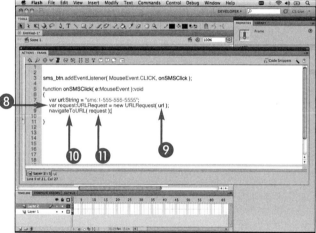

Extra

As well as being launched from your applications, the sms URL can be used as an HTML link on a Web page in the Safari application on your device. You can also include an sms URL in HTML text that is rendered in a TextField. To apply HTML text to your TextFields, you must first select the Render As HTML button in the Properties panel of the TextField. After you select this, you can set the htmlText property of your TextField to the HTML text String:

```
var html:String = '<u><a href="sms:+1-555-555-5555">Send me a text message</a></u>';
mytext_txt.htmlText = html;
```

Play a YouTube Video

The YouTube URL protocol can be used to play videos in the YouTube application. If the YouTube application is not available or found, the Safari application on the device will open to the YouTube Web site. Similar to linking to the Maps application, the YouTube URL protocol does not use a `youtube` scheme identifier. Instead, it uses a regular `http` protocol, which points to the YouTube servers.

Before you can link to a specific YouTube video, you must first get the unique identifier for that video. The unique identifier for a video can be found at the end of the URL in the Address bar when on the video page of the YouTube site. For example, if the URL to your video was http://www.youtube.com/watch?v=Nr32TcO7fmM, the unique video identifier would be Nr32TcO7fmM.

There are two different ways you can format the link to a YouTube video. The first is similar to the URL you would see in the Address bar if you went to the video in a browser on your desktop — http://www.youtube.com/watch?v=VIDEO_IDENTIFIER, in which VIDEO_IDENTIFIER is the identifier of your video. When in a browser, this URL will bring you to the page for your video on YouTube.

The second URL format is slightly different from the first — http://www.youtube.com/v/VIDEO_IDENTIFIER, in which VIDEO_IDENTIFIER is the identifier of your video. If you went to this URL in a browser, it would display the video full screen outside the shell of the YouTube Web site.

When navigated to, both URL formats will exit your application and launch the YouTube application. If the video can be played on the device, it will automatically be launched and start to play.

Play a YouTube Video

1. In a Web browser, navigate to the YouTube video page.

2. Copy the video identifier.

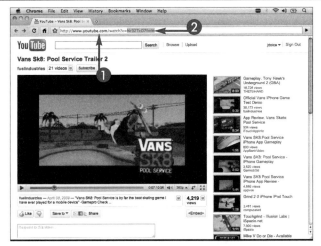

3. In Flash, create a `Button` symbol and place it on the Stage.

Note: See Chapter 2 for more information.

4. Give it an instance name, such as `youtube_btn`.

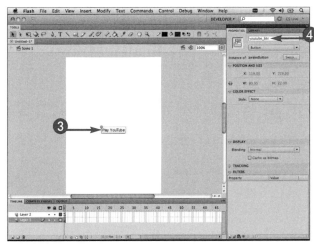

5. Register a listener for the button click, such as `youtube_btn.addEventListener ( MouseEvent.CLICK, onYBClick );`.

6. Create an event handler function, such as `onYBClick`.

7. Create a String variable, such as `var url:String = "";`.

8. Add the YouTube URL, such as `http://www.youtube.com/v/`.

9. Add the video identifier, such as `Nr32TcO7fmM`.

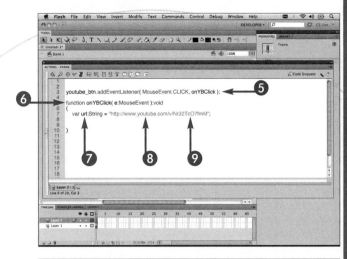

10. Create a URLRequest variable, such as `var request:URLRequest = new URLRequest();`.

11. Add the URL to the request.

12. Type **navigateToURL();**.

13. Add the `request` variable as an argument.

After you compile and install the application on a device, when the button is pressed, the YouTube application will launch, and the video will begin to play.

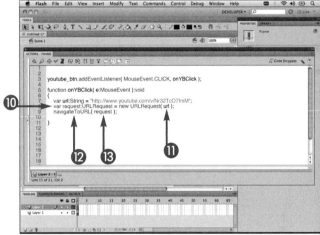

Extra

It is important to note that not all YouTube videos can be played on the iPhone. If you link to a video that is not supported for playback on the device, you will be brought to the YouTube video page within the application, and an error message will be shown. Videos that were uploaded to YouTube before the iPhone was released are the ones that may not be supported on the device. This is because they were not encoded in a format in which the iPhone understands. Since the release of the iPhone, YouTube has started to encode its videos using the H.264 video codec, which is a supported video codec for iOS. YouTube has begun re-encoding videos that were uploaded and not encoded with the H.264 video format, in order to support the iPhone and other devices. However, there is the possibility that your video did not get re-encoded yet, so you should make sure to test all your videos before submitting your application. If you have recently uploaded your video to YouTube, it will most likely support playback on the device.

Open the iTunes Store

Opening the iTunes Store to a specific item gives your users the ability to purchase and download music and other items in the store. The URL scheme for the iTunes Store is very complicated to construct. Therefore, in order to help in creating the URL, Apple has created an online tool called the *iTunes Link Maker*. This tool enables you to select a media type, such as music, movies, TV shows, and podcasts, and the country of the store that you want to link to. You can then input a search term into the text box to bring up a list of the available titles. The Search box acts similarly to the Search box in the iTunes application.

After you have entered your search terms, a list of all the results will be shown. All the appropriate information will be shown for the type of media that you searched for. For example, when you are searching for music, the song name, album name, and artist name will all be shown. For each piece of information that can be linked to, a gray circle with a white arrow will be shown beside it. Clicking this button will bring up the generated HTML link for that item. Simply copy the direct URL contained inside the HTML text and place it inside your application.

Alternatively, you can navigate to the item in the iTunes Store within the iTunes application and right-click an item to copy the link. For example, if you wanted to copy the link of your application, navigate to the application page and right-click its title.

Open the iTunes Store

1. In a Web browser, navigate to the iTunes Link Maker, at www.apple.com/itunes/linkmaker/.

2. Click here and select a country, such as Canada.

3. Click here and select a media type, such as Music.

4. Enter a search term, such as Norah Jones.

5. Click Search.

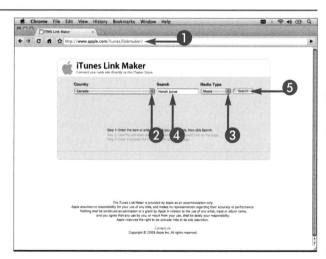

6. Click the button for the item that you want to link to, such as The Fall (Deluxe Version).

⑦ Copy the URL in the HTML text box, such as http://itunes.apple.com/ca/album/chasing-pirates/id337904641?i=337904941&uo=6.

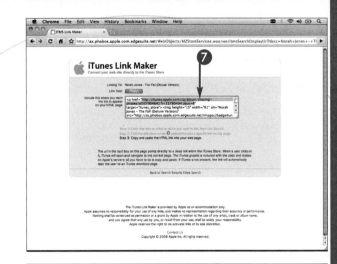

⑧ In Flash, open the Actions panel.

⑨ Create a `String` variable with the URL, such as `var url:String = "http://itunes.apple.com/ca/album/chasing-pirates/id337904641?i=337904941&uo=6";`.

⑩ Create a `URLRequest` variable, such as `var request:URLRequest = new URLRequest();`.

⑪ Add the URL to the request.

⑫ Type **navigateToURL();**.

⑬ Add the `request` variable as an argument.

After you compile and install the application on a device, when the button is pressed, the iTunes application will open to the specified item.

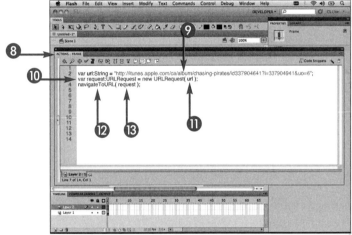

Extra

If you plan on linking to items within the store, it is a good idea to join the iTunes Affiliate Program. The iTunes Affiliate Program allows you to place links on your Web site to the U.S. iTunes Store and receive commissions based on qualified purchases. This is especially great if you plan on promoting your application on your own Web site. LinkShare, a leading Internet provider of innovative technology that helps affiliates sell more products and earn more commissions, administers the Affiliate Program. After you have signed up for a LinkShare account, you can apply to become an iTunes Affiliate member. When you are approved, you will be able to access all the tools in order to create links for the specific items in the iTunes Store.

For each qualified sale generated by links to iTunes on your Web site, you will receive a 5 percent commission. There are affiliate programs set up for Europe, Japan, Australia, and New Zealand, as well as the United States. However, if you live outside of these markets, you can still become an affiliate. For more detailed information about the iTunes Affiliate Program, go to its Web site at www.apple.com/itunes/affiliates/.

Submit Updates to Twitter

The iPhone and iPod touch have become the premier social devices on the market today. The ability to keep in touch with your friends and contacts wherever you are is a key element to the success of any new mobile device. Games and applications are taking advantage of this by integrating social media into them. Many games in the App Store will enable you to post your score to your Twitter feed. Integrating Twitter and other social media touch points gives developers free advertising within their users' social media networks.

Submitting text to users' Twitter feeds can be done using the Twitter API. The `twitter.com/statuses/update.xml` URL is used to POST the status string. You will need to have a method for the users to input their Twitter usernames and passwords. It is a good idea to store these locally so that the users do not have to input them every time. The username and password get added to the URL, creating the URL `http://username:password@twitter.com/statuses/update.xml`.

The status query string parameter is used to set the status update text. You can use a `URLVariables` instance to store this variable. Your `URLVariables` instance is then set to the `data` property of your `URLRequest` instance for the URL.

To send the Twitter status update, use a `URLLoader` instance. This allows you to listen for result events to be fired when the URL has been submitted. Registering a listener for the `IOErrorEvent.IO_ERROR` event enables you to notify the users that there was an error in submitting their Twitter status. This could be caused by an incorrect username or password, or there may be issues with the Twitter API. If an `Event.COMPLETE` event is fired, the submission was successful.

Submit Updates to Twitter

1. Create a `String` variable, such as `var url:String = "";`.

2. Add the Twitter status update URL, such as `http://@twitter.com/statuses/update.xml`.

3. Add the username.

4. Add the password.

5. Create a `URLRequest` variable with the URL, such as `var request:URLRequest = new URLRequest(url);`.

6. Create a `URLVariables` variable, such as `var variables:URLVariables = new URLVariables();`.

7. Set the status text, such as `variables.status = "I just scored 1000 in Sushi Toss";`.

8. Set the variables to the request, such as `request.data = variables;`.

9. Set the request method to POST, such as `request.method = URLRequestMethod.POST;`.

10. Create a new `URLLoader` variable, such as `var loader:URLLoader = new URLLoader();`.

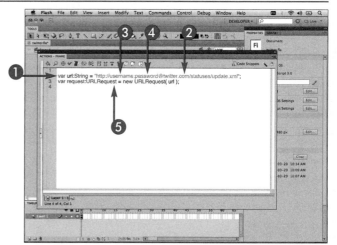

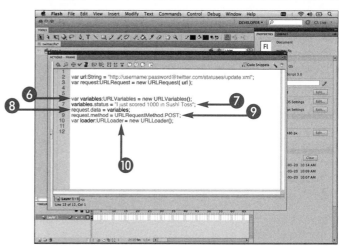

⑪ Listen for any errors, such as `loader.addEventListener(IOError Event.IO_ERROR, ioError );`.

⑫ Create an event handler function, such as `ioError`.

⑬ Output any errors, such as `trace( e.toString() );`.

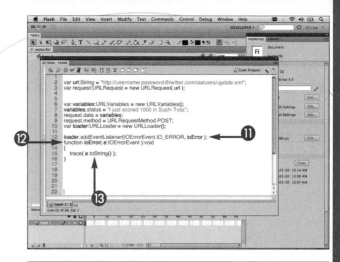

⑭ Listen for any complete events, such as `loader.addEventListener (Event.COMPLETE, onComplete );`.

⑮ Create an event handler function, such as `onComplete`.

⑯ Output the result data, such as `trace( loader.data as String );`.

⑰ Load the request, such as `loader.load( request );`.

The status update will be posted to Twitter.

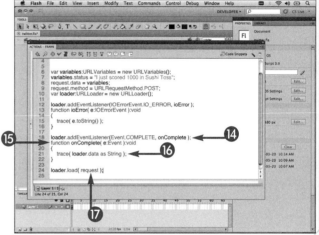

Apply It

Twitter has exposed an entire set of API methods for you to integrate with. If you want to integrate more of Twitter's features in your application, you can use one of the many open source ActionScript Twitter libraries. My favorite is `twitterscript`, available at http://code.google.com/p/twitterscript/, which was originally developed by Twitter and later open sourced in order for it to be maintained and kept current with API changes. The following is an example of how to use the `twitterscript` API to load the home Timeline of a user:

```
var twitter:Twitter = new Twitter();
twitter.setAuthenticationCredentials(username, password );
twitter.addEventListener(TwitterEvent.ON_HOME_TIMELINE_RESULT, onHomeResult );
twitter.loadHomeTimeline();
private function onHomeResult(event:TwitterEvent):void {
for( var i:int = 0; i<event.data.length; i++ ){
var status:TwitterStatus = event.data[ i ] as TwitterStatus;
trace( status.text );
}
}
```

Display Ads with Smaato

If you decide to create a free application, you can add advertising to your application in order to receive some earnings. Many of the free applications in the store use this method to allow users to download a free version of their application and turn off the ads if they purchase the full app. Smaato is one of the leading mobile advertising companies on the market today. Its API enables you to place ads inside your application as images or as text. To get started integrating ads in your application, go to the Smaato Web site at www.smaato.com and register for an account.

After you have registered for an account and set up your application, you will be given a `PublisherID` and an `AdSpaceID`. Keep these on hand because you will need them when constructing the ad request. You can also use 0 for both IDs if you want to test the API without registering your application.

When making an ad request, you can specify in which format the results will be returned. The easiest format to work with is XML. Having the result data in XML enables you to easily parse the image URL and the URL to navigate to when the user taps the image. After the URL of the image has been parsed, a second request will need to be made in order to load the image into the banner. This can be done using the `Loader` class. For more details on loading external images, see Chapter 6, "Working with Images."

When designing your application, you will want to account for a banner dimension size of 300 x 50.

Display Ads with Smaato

1. Create a `URLRequest` variable with the Smaato API URL, such as `var request:URLRequest = new URLRequest( "http://soma.smaato.com/oapi/reqAd.jsp" );`.

2. Create a `URLVariables` variable, such as `var variables:URLVariables = new URLVariables();`.

3. Add your `AdSpaceID`, such as `variables.adspace = "0";`.

4. Add your `PublisherID`, such as `variables.pub = "0";`.

5. Set the local IP, such as `variables.devip = "127.0.0.1";`.

6. Set the ad format, such as `variables.format = "IMG";`.

7. Set the number of ads to return, such as `variables.adcount = 1;`.

8. Set the response data format, such as `variables.response = "XML";`.

9. Add the variable to the request, such as `request.data = variables;`.

10. Create a `URLLoader` variable, such as `var loader:URLLoader = new URLLoader();`.

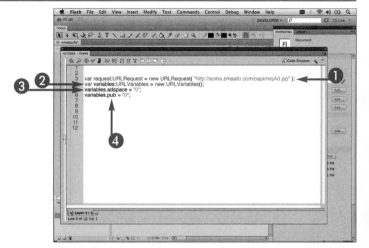

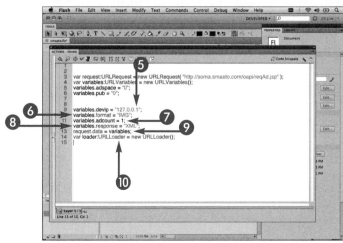

⑪ Listen for the complete event, such as `loader.addEventListener( Event.COMPLETE, onComplete );`.

⑫ Load the request, such as `loader.load( request );`.

⑬ Create an event handler function.

⑭ Parse the data as XML, such as `var data:XML = new XML(loader.data as String);`.

⑮ Parse the status, such as `var status:String = data.*::status.toString();`.

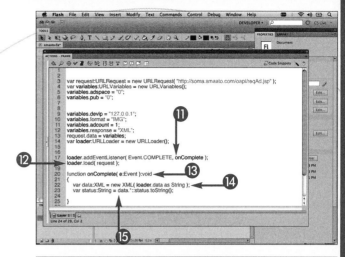

⑯ Check the status, such as `if( status == "success" ){}`.

⑰ Parse the ad, such as `var ad:XMLList = data.*::ads.*::ad;`.

⑱ Parse the image URL, such as `var link:String = ad.*::link.toString();`.

⑲ Create a Loader variable, such as `var l:Loader = new Loader();`.

⑳ Load the image, such as `l.load( new URLRequest( link ) );`.

㉑ Add the image to the display list, such as `addChild( l );`.

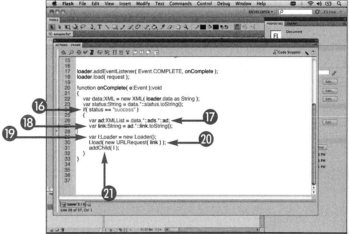

The ad will be displayed in your application.

Apply It

After the image for the ad has been loaded, you can register it to listen for a `MouseEvent.CLICK` event in order to navigate to the URL of the ad. The following is a continuation of the example to add this functionality:

```
addChild( l );
var clickurl:String = ad.*::action.@target.toString();
l.addEventListener( MouseEvent.CLICK, onAdClick );
function onAdClick( e:MouseEvent ):void{
var request:URLRequest = new URLRequest( clickurl );
navigateToURL( request );
}
```

Track with Google Analytics

Integrating a form of analytics in your application is important in understanding how your users are using your application. Checking how many people have downloaded your application is only part of understanding how popular your application is. It allows you to see which portions of your application users are using more than others. It also enables you to determine if there are any features in your application that users are not using.

One of the more popular analytics suites is Google Analytics. Google Analytics provides a great feature set and is free to use. When you set up your application with Google Analytics, a unique ID will be generated for your application. You will want to copy your ID, as you will use it in your application. Google has also developed an open source ActionScript 3.0 API, which you can use to help with tracking interactions within your application.

The `GATracker` class contains all the methods to send tracking calls to Google Analytics. The constructor has two required parameters: The first parameter is the main Stage of your application. The `GATracker` uses the Stage or any another `DisplayObject` in the display list for debugging purposes. The second parameter is a `String` variable, which is the Google Analytics unique ID for your application.

After you have instantiated your `GATracker`, you can send tracking calls to the Google Analytics servers. The most common item to track is page views. Whenever your application switches its view or the user navigates to a new section, you will want to call the `trackPageview();` method on the `GATracker` instance, with a `String` representing the section as its parameter.

Track with Google Analytics

1. In a Web browser, sign up for a Google Analytics account at www.google.com/analytics/.

2. Log into your Google Analytics account.

   The Google Analytics home page appears.

3. Click Add Website Profile.

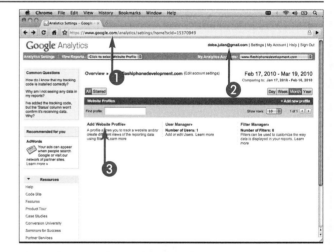

   The Create New Website Profile page appears.

4. Enter a URL to track against.

   Note: *This will act as the domain for your application.*

5. Click here and select your country.

6. Click here and select your time zone.

7. Click Finish.

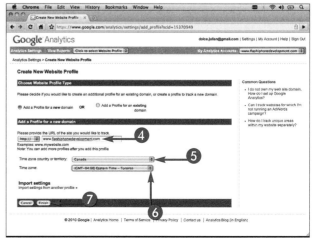

The Tracking Code page appears.

8 Select and copy the Web property ID.

Note: You can download the Google API from the Google Code project at http://code.google.com/p/gaforflash/.

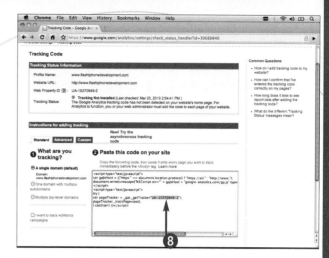

9 In the Actions panel in Flash, import the necessary Google Analytics classes.

10 Create a GATracker variable, such as `var tracker:AnalyticsTracker = new GATracker();`.

11 Pass in the Stage as the first parameter.

12 Enter your Web property ID as the second parameter.

13 Track a page view, such as `tracker.trackPageview( "/home");`.

The tracking request will be sent to Google.

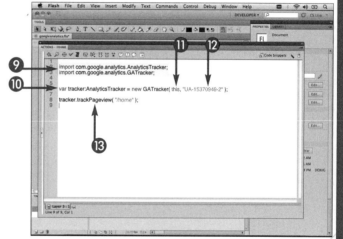

Apply It

As well as track page views, you can track events. Events can be anything that occurs in your application. Selecting a button, completing a level in a game, and successfully logging in are all examples of events that can be tracked. To track your event, call the `trackEvent()` method on the GATracker instance. The method takes four parameters: The first parameter is the category for the event. The second parameter is the action that created the event. The third parameter is an optional descriptor for the event. The last parameter is an optional value to be aggregated with the event. The following is an example of tracking a button click as an event:

```
var tracker:AnalyticsTracker = new GATracker( this, "GA-ID" );
btn.addEventListener( MouseEvent.CLICK, onButtonClick );
function onButtonClick( e:Event ):void{
tracker.trackEvent( "Button", "click", "Button", 123 );
}
```

Display Ads with AdMob

AdMob is one of the premier mobile advertising platforms on the market today. AdMob provides a variety of SDKs, which you can use to distribute and monetize your applications. Currently, there is not a Flash for iPhone SDK; however, AdMob has just released a beta version of the Flash Lite SDK. The SDK is completely written in ActionScript 2.0 and will not work with the Flash iPhone Packager, as all applications must be written in ActionScript 3.0.

Having the ActionScript 2.0 source code for the Flash Lite SDK, however, allows us to convert it to ActionScript 3.0. The ads that are served with the Flash Lite version are 192 pixels wide and 53 pixels high, which is not the optimal size for the iPhone. Having full control over the ad placement, you can be creative and place it in a location that will still look good.

After you sign up for an AdMob account, you can register a Flash Lite site. For each site that you create, a site ID will be generated for you. This ID is passed into the constructor of the AdMob class, which is included in the ad request.

The second parameter for the AdMob class constructor is the target holder for the ad that is returned. When the ad is loaded, it is placed at the top-left corner; or 0,0; of the holder. Knowing the width of the ad and the width of the Stage, you can easily center the holder so that the ad is centered on the Stage after it has been loaded.

The third parameter is an optional Boolean, which specifies to set the ad request in test mode. This allows you to test the ads during development and turn on the live ads only when your application has been published to the App Store. The default value for this parameter is false.

Display Ads with AdMob

1. Go to AdMob's Web site, www.admob.com, and create an account.

2. Obtain an AdMob site ID.

3. In the Actions panel in Flash, create a Sprite variable, such as var adHolder:Sprite = new Sprite();.

4. Set the x property of the Sprite, such as adHolder.x = .

5. Get the center of the Stage width, such as (stage.stageWidth/2).

6. Subtract half the width of the ad, such as -(192/2);.

7. Create an AdMob variable, such as var admob:AdMob = new AdMob();.

8. Enter your site ID.

9. Set the holder target for the ad.

10. Submit ad requests in test mode, such as true.

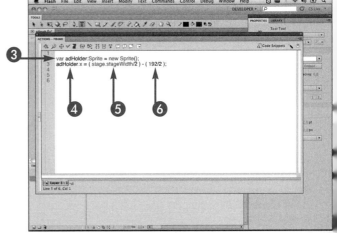

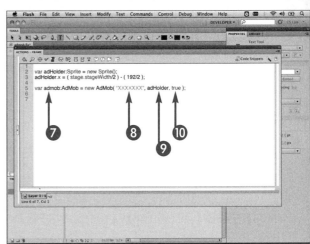

⓫ Add the ad container to the Stage, such as `addChild( adHolder );`.

⓬ Load the ad, such as `admob.loadAd();`.

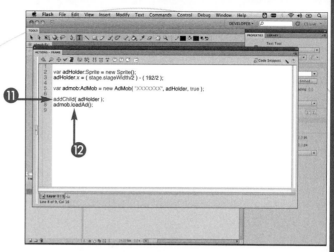

⓭ Compile the application.

● The ad is centered at the top of the application.

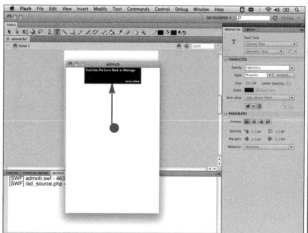

Extra

When you register your site with AdMob, you have the option to set the colors of your ad when it is displayed in your application. These can be changed anytime from the App Settings tab in the Settings for your site. The default ad style is a black background with white text. To change the colors, click the App Settings tab in the Settings section for your site. There are two options for the ad style: By default, Use Colors Set in Client Mode is selected. Selecting Use Colors Set Below will allow you to customize the colors. There is a text input box for the text color and background color in which you can enter the hex color values that you want. You can also select the color wheel icon beside each text box in order to show a list of predefined colors.

With your custom colors set for the ads that are returned, you will have more control over how they are integrated into the design of your application. One idea is to draw a colored bar across the entire width of the Stage the same color as the background of the ad.

Optimize Your Display List

One of the biggest bottlenecks when it comes to performance is the rendering of the display list. In comparison, your ActionScript code will outperform the rendering of your visual assets. So even if you have taken every step in order to make sure that your code is optimized to be as fast as possible, your visual assets may still be slowing your application down. There are several things to watch out for when creating your visual assets that will help you to make sure that your application will perform as fast as possible.

One tip is to make sure that you keep the display list as shallow as possible. Try not to have too many `DisplayObject`s with many nested clips. When a touch event occurs on the screen, the display list is traversed to find all the elements that are underneath the user's finger. The deeper the display list, the longer the traversal will take. Keeping the list shallow will return those objects faster.

Another tip is to reduce the amount of items that overlap each other. The more items that are overlapping, the more compositing has to occur. In addition, try to reduce the amount of alpha in your images. The more alpha there is in your image, the more the renderer has to composite what is underneath the alpha.

My third tip is to remove all your masks. Masking items will cause them to be redrawn to the screen every frame, even when a redraw is not necessary. This will cause an extreme decrease in performance as your application will render more than it needs to.

The example in the steps below shows how you could organize your visual assets in order to create a scrollable list.

Optimize Your Display List

Optimize a Scrollable List

1. Create a `MovieClip` symbol to represent a cell in the list.

Note: See Chapter 2, "Getting Started with Flash CS5," for more information.

2. Stack the cells on the Stage exceeding the height of the Stage.

Note: Make sure that the cells do not overlap and that they are on a whole pixel.

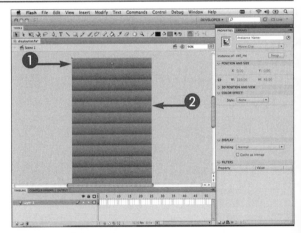

3. Click the Selection tool.
4. Select all the cells on the Stage.

Note: You can also press the ⌘+A (Ctrl+A) keyboard shortcut to select them all.

5. Convert the selected cells to a `MovieClip` symbol by pressing the F8 key.

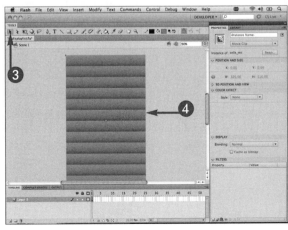

6 Create a new layer on the Timeline above the layer with the cells.

7 Click here to hide the layer with the cells on it.

8 Place a header graphic at the top of the Stage.

9 Place a footer graphic at the bottom of the Stage.

Note: Covering objects are a way to reduce the number of masks needed.

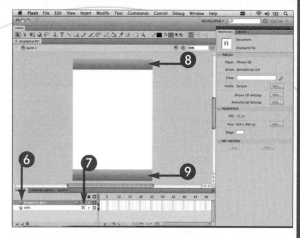

Test Your Movie

10 Press ⌘+ Enter (Ctrl+Enter).

- The extra cells are covered by the header and footer graphics.

  Cells are not overlapped or masked, which improves performance.

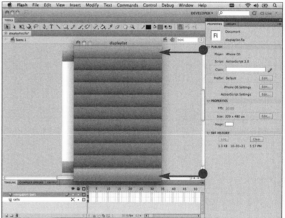

Chapter 15: Optimizing Performance

Extra

Using vectors can have a big impact on the performance of your application. A good practice to follow is to eliminate the use of vector objects unless absolutely necessary. This may not always be possible — for example, if you are creating a paint application. In these instances, you may be able to get a good enough performance if the application is not that intensive. However, if you are creating a game and you have some illustrated characters and objects in vector, you will want to export them to images. The amount of detail and complex vector points that an animated character can have is so great that the renderer will have a hard time keeping up with a decent frame rate.

From within the Flash IDE, you can easily export each frame as a PNG or other image format. Reimporting your vector assets as images will greatly improve your chances of achieving a decent frame rate. Exporting a lot of frames from a Timeline animation can be a very time-consuming process. You can automate this by creating a JSFL script file that will move the play head to each frame and export it to an image.

Manage Mouse Events

As mentioned in the preceding section, "Optimize Your Display List," keeping your display list shallow will have an impact on the performance of your application. One of the reasons is how mouse and touch events are handled within ActionScript 3.0.

There are three phases to a MouseEvent. The capturing phase is the first phase and occurs when an event is fired. The event starts with the topmost parent display object, or the Stage, and works its way down the display list hierarchy until it reaches the target in which the event originated. The second phase is the target phase, which occurs when the event reaches the display object from which the event originated. The third phase is the bubbling phase. During the bubbling phase of an event, the event follows the reverse path of the capturing event, all the way to the topmost parent display object.

With this knowledge, you can see how having deeply nested display lists can easily and quickly cause a decrease in performance, as the event will visit each item in the display list in both the capturing and bubbling phases.

There are a couple methods on the Event class that enable you to stop the event in its current phase. The stopPropagation() method prevents the processing of any event listeners in any nodes subsequent to the current node in the event flow. Calling this method will not have any effect on the current node, and it can be called during any phase of the event.

The stopImmediatePropagation() method prevents the processing of any event listeners in the current and subsequent nodes in the event flow. This method takes effect immediately and affects event listeners in the current node. However, you should note that this does not cancel the behavior of the associated event.

Manage Mouse Events

① Add a Button symbol to the Stage.

Note: *See Chapter 2 for more information.*

② Give it an instance name, such as btn.

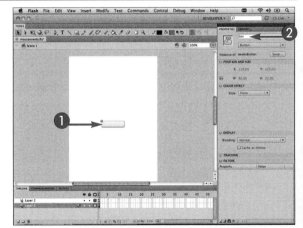

③ In the Actions panel, add a CLICK listener to the Button, such as btn.addEventListener(MouseEvent.CLICK, onClick);.

④ Create an event handler function.

⑤ Output the phase of the event, such as trace("button clicked", e.eventPhase);.

⑥ Stop the event from bubbling up, such as e.stopImmediatePropagation();.

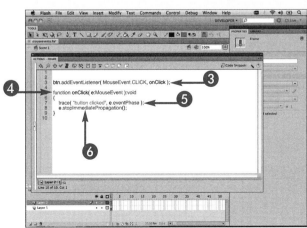

7 Listen for the bubbled event on the Stage, such as `stage.addEventListener(MouseEvent.CLICK, onStageClick );`.

8 Create an event handler function.

9 Output the phase of the event, such as `trace("stage clicked", e.eventPhase);`.

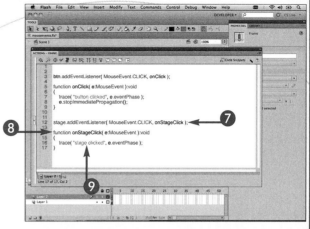

10 Press ⌘+ Enter (Ctrl+Enter) to test your movie.

● Check the Output panel to see the event phases.

The propagation of the event has been stopped and will not bubble.

Note: *You can remove the* `e.stopImmediatePropagation();` *line of code in order to see the event bubble up to the Stage.*

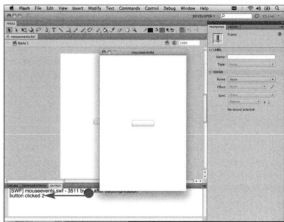

Chapter 15: Optimizing Performance

Apply It

When the user drags his or her finger while it is pressed on the screen, a `MouseEvent.MOUSE_MOVE` event is fired. This event can be fired multiple times a frame as the finger moves quickly across the screen, which can decrease performance. The following is an example of how to use the `Event.ENTER_FRAME` event to track the position of a touch:

```
stage.addEventListener( Event.ENTER_FRAME, onFrame );
function onFrame( e:Event ):void{
var x:Number = stage.mouseX;
var y:Number = stage.mouseY;
trace( x, y );
}
```

Understanding cacheAsBitmap

The `cacheAsBitmap` property is a property that exists on all `DisplayObject` instances. If the `cacheAsBitmap` property is set to `true`, a bitmap representation of the display object will be cached internally. To get the highest performance increases in your application, use this method in conjunction with setting the rendering mode to GPU. This can be done in the iPhone Settings dialog box.

When your display object is cached using the `cacheAsBitmap` property, it is sent to the GPU. This enables you to perform simple translations along the x and y axes without having the display object reuploaded to the GPU. This works very well with display objects whose states remain fairly constant.

If there are any changes to the bounds of the display object — for example, if the object is scaled or rotated, the bitmap will be re-created and uploaded to the GPU. For more information on scaling and rotating display objects, see the following section, "Understanding cacheAsBitmapMatrix."

The `cacheAsBitmap` property is automatically set to `true` when a filter is applied to the display object. If you set `cacheAsBitmap` to `false` and then apply a filter, it will set the property to `true`. After all filters are removed from the display object or the `filter` array is empty, the property will be reset to its previous value.

When the `cacheAsBitmap` property is set to `true`, the rendering does not change. However, the display object snaps to the nearest pixel. To ensure that you do not see a visual shift in your display objects when setting the `cacheAsBitmap` property, make sure that all your elements are on whole pixels.

Understanding cacheAsBitmap

① Create a `MovieClip` symbol and place it on the Stage.

Note: See Chapter 2 for more information.

② Give it an instance name, such as `sushi_mc`.

Note: You can set the `cacheAsBitmap` property in the IDE as well as in code.

③ Click Cache As Bitmap.

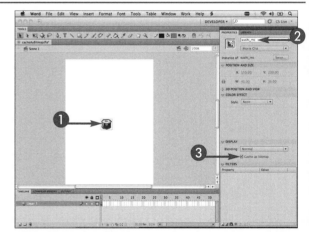

④ In the Actions panel, cache the symbol, such as `sushi_mc.cacheAsBitmap = true;`.

⑤ Register a listener for the enter frame event, such as `stage.addEventListener( Event.ENTER_FRAME, onFrame );`.

⑥ Create an event handler.

⑦ Move the symbol incrementally every frame, such as `sushi_mc.y -= 2;`.

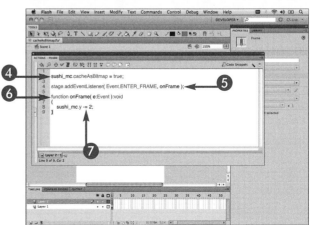

8 Click File.

9 Click iPhone OS Settings.

The iPhone Settings dialog box appears.

10 Click here and select GPU.

11 Click OK.

When you compile and install your application on your device, the `cacheAsBitmap` property will upload a representation of the display object to the GPU.

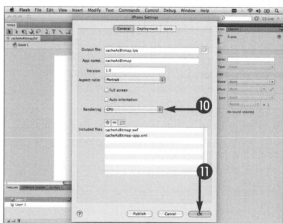

Extra

The `opaqueBackground` property on a `DisplayObject` instance specifies whether it is opaque and the color value of the background. Setting this property to a number value creates an opaque background on the object to the color value that the value specifies. If it is set to null, the background of the display object will be transparent. When the `cacheAsBitmap` property is set to `true`, setting the display object to opaque can improve rendering performance. It is important to note that the opaque background region does not respond to mouse events.

```
myshape.opaqueBackground = 0x000000;
myshape.cacheAsBitmap = true;
```

Chapter 15: Optimizing Performance

Understanding cacheAsBitmapMatrix

The preceding section, "Understanding cacheAsBitmap," explores how to use the `cacheAsBitmap` property to upload display objects to the GPU. This works well if the bounds of the display object do not change, such as if it is not scaled or rotated. If your display objects are required to be scaled or rotated, however, you can use the `cacheAsBitmapMatrix` property to prevent the bitmap from being re-created and uploaded to the GPU.

Setting the `cacheAsBitmapMatrix` property to a valid `Matrix` instance will define how the display object is rendered when `cacheAsBitmap` is set to `true`. Your application will use this `Matrix` as a transformation matrix when rendering the bitmap version of the display object.

When the `cacheAsBitmapMatrix` property is set, the application will retain a cached version of the bitmap when a transformation of the display object occurs, such as rotation, scale, and translation. When the rendering mode of the application is set to GPU, the display object will be stored as texture in video memory. This allows all supported transformation to occur on the GPU, which can perform these transformations much faster than the CPU.

Simply setting the property to the identity matrix — for example, `new Matrix();` — usually suffices, however, if you use any `Matrix` instance to upload a different bitmap to the GPU. It is best practice to set the `Matrix` to the size that it will appear in the application.

Transformations of the display object are not required to occur on the matrix. After the `cacheAsBitmapMatrix` property is set, you can use the rotation and scale properties of the display object in order to transform the object.

Understanding cacheAsBitmapMatrix

1. Create a `MovieClip` symbol and place it on the Stage.

Note: See Chapter 2 for more information.

2. Give it an instance name, such as `sushi_mc`.

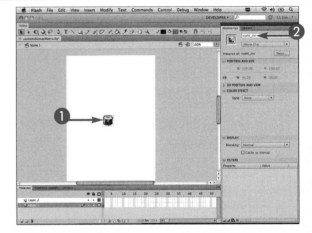

3. In the Actions panel, cache the symbol, such as `sushi_mc.cacheAsBitmap = true;`.

4. Cache the matrix, such as `sushi_mc.cacheAsBitmapMatrix = new Matrix();`.

5. Register a listener for the enter frame event, such as `stage.addEventListener( Event.ENTER_FRAME, onFrame );`.

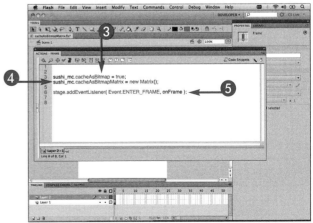

Chapter 15: Optimizing Performance

6. Create an event handler.

7. Move the symbol incrementally every frame, such as `sushi_mc.y -= 2;`.

8. Rotate the symbol incrementally every frame, such as `sushi_mc.rotation += 10;`.

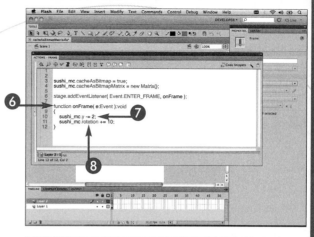

9. Click File → iPhone OS Settings.

    The iPhone Settings dialog box appears.

10. Click here and select GPU.

11. Click OK.

    When you compile and install your application on your device, the `cacheAsBitmapMatrix` property will upload a representation of the display object to the GPU.

    Transformations to the display object will not cause it to be redrawn.

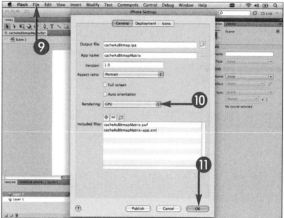

Extra

It is important to understand that the `cacheAsBitmapMatrix` property works only on 2D surfaces, as the texture that is uploaded is only a 2D surface. If you want to use a 3D surface, you simply set the z property to 0 or the value that you want. Setting the z property automatically sets your display object to be a 3D surface. The following syntax sets a Sprite instance to be a 3D surface:

```
mysprite.z = 0;
```

There is also an easy way to set the `cacheAsBitmapMatrix` to the same Matrix as it currently is on the Stage:

```
mysprite.cacheAsBitmapMatrix = sprite.transform.matrix.clone();
```

This is very useful when your display object has already been transformed before you set the matrix.

Determine the Device OS

One of the hardest things about mobile development for a specific platform is trying to target all the different devices. There are a number of devices that ship with Flash Lite; however, a lot of the phones have different screen resolutions and capabilities. Even the BlackBerry has at least four different screens that you need to build for.

One of the nice things about the iOS platform has been the consistency of the devices. When the iPhone SDK first launched, it was very easy to test your applications, as you had to test only on one or two different devices. But as Apple releases better and faster iPhone and iPod touch versions, it will become increasingly more difficult to target all the various devices — not to mention that the iPad is set to change everything.

Knowing the device type and its generation will become increasingly important as iOS devices become faster. Currently, there is a big difference in performance between the iPhone 3G and the 3GS. Some games simply do not have a decent performance on the 3G. Determining the device will allow you to make adjustments to your application for the various devices.

For example, on the slower devices, such as the first- and second-generation iPhones and iPod touches, you may want to remove functionality or reduce the amount of content on the screen at a given time to improve performance, whereas you can provide the users of the iPhone 3GS with the full experience.

The Capabilities.os property returns the device model as well as the system version as one string. These are the same values that are returned when using the UIDevice.model and UIDevice.systemVersion in the iPhone SDK.

Determine the Device OS

1. Click the Text tool.
2. Create a text field on the Stage.
3. Click here and select Classic Text.
4. Click here and select Dynamic Text.
5. Give it an instance name, such as device_txt.

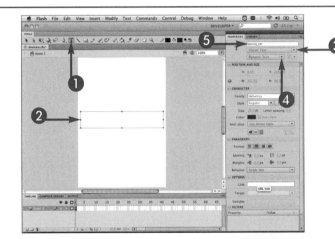

6. Get the OS type, such as var os:String = Capabilities.os;.
7. Check for a first-generation iPhone.
8. Set the text, such as device_txt.text = "first gen iPhone";.

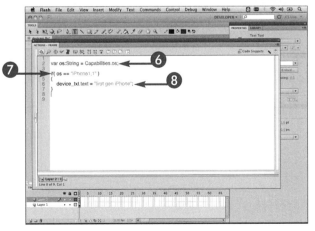

⑨ Check for the iPhone 3G.

⑩ Set the text, such as `device_txt.text = "iPhone 3G";`.

⑪ Check for the iPhone 3GS.

⑫ Set the text, such as `device_txt.text = "iPhone 3GS";`.

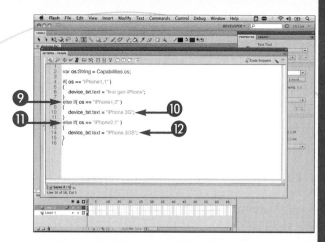

⑬ Check for the first-generation iPod touch.

⑭ Set the text, such as `device_txt.text = "first gen iPod Touch";`.

⑮ Check for the second-generation iPod touch.

⑯ Set the text, such as `device_txt.text = "second gen iPod Touch";`.

When you compile and install your application on your device, which device is being used will appear in the text field.

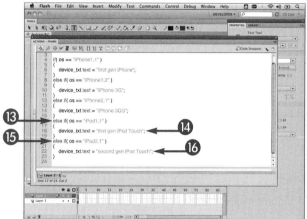

Apply It

Being able to know the device OS also enables you to check which platform you are running on while testing locally. For example, if you are building an application that makes use of the accelerometer, you will always have to upload the application to the device, as there is no accelerometer support in the Flash IDE. To work around this limitation, you could check the `Capabilities.os` property and determine if the application is running locally. If it is, you could then show a button or provide some other functionality to simulate the accelerometer. This will save you from constantly uploading the application to your device, allowing you to spend more time developing and not waiting. The following syntax will check to see if the application is running locally:

```
var os:String = Capabilities.os;
if( os.indexOf( "Mac" ) != -1 || os.indexOf( "Windows" ) != -1 ){
trace( "running locally" );
}
```

Create a Settings Bundle

In most desktop application scenarios, application settings are stored and set within the application itself. iOS takes a slightly different approach and provides a Settings application that handles any custom settings for your application. This is completely optional and enables you to specify a custom user interface for the preferences.

To integrate your custom application settings into the Settings application, you create a Settings.bundle file and bundle it in the root of your application directory. This file is responsible for determining the layout, types of inputs, and default values for your settings. When your application launches, you are able to retrieve these settings at runtime with code, which is explored later in this chapter.

The easiest way to create a Settings.bundle file is in Xcode if you are developing on a Mac OS X system. The Settings.bundle file is actually a folder that contains a Root.plist file, which describes the settings interface. Also, it has a strings folder for each language that your application is localized in. This allows the labels for the controls in the Settings application to retrieve the proper strings based on which language the device is currently set to.

To create a Settings.bundle file in Xcode, start a new iPhone project. You can pick any of the iOS templates, as you just want to be able to create the Settings.bundle file. After your project has been created, you use the New File dialog box to create the Settings.bundle file. Xcode will prefill a few default settings for you, which are easily removed in order to start fresh.

Create a Settings Bundle

Note: *These steps work only on Mac OS X. If you are using Windows, see the Apply It on the next page.*

1. In Xcode, click File → New Project.

   The New Project dialog box appears.

2. Click Application under the iPhone OS header.

3. Select a project template, such as Utility Application.

4. Click Choose.

   A file dialog box appears.

5. Enter a name for your project.

6. Click Save.

   Your project is created.

7. Click File.

8. Click New File.

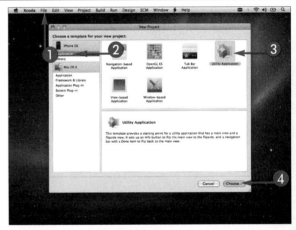

The New File dialog box appears.

9 Click Resource under the iPhone OS header.

10 Click Settings Bundle.

11 Click Next.

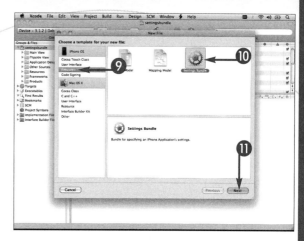

The New Settings Bundle page of the dialog box appears.

12 Name the file Settings.bundle.

13 Click Finish.

Your Settings.bundle file is created.

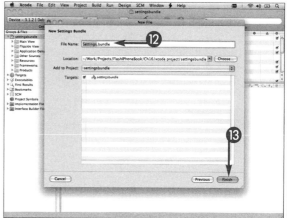

Apply It

If you are developing on a PC, this process is a little more difficult. Because all the files are simply text-based XML, you can create them yourself. First, create a folder called Settings.bundle. Inside that folder, create a text file called Root.plist and copy the following into it:

```xml
<?xml version="1.0" encoding="UTF-8"?>
<!DOCTYPE plist PUBLIC "-//Apple//DTD PLIST 1.0//EN" "http://www.apple.com/DTDs/
   PropertyList-1.0.dtd">
<plist version="1.0">
<dict>
                <key>StringsTable</key>
                <string>Root</string>
                <key>PreferenceSpecifiers</key>
                <array>
                </array>
</dict>
</plist>
```

Add a Settings Group

Before you start building your Settings.bundle file, it is a good idea to plan it out first. Look at the Settings preferences' interfaces for some of your favorite applications on your device. Planning your preferences out on paper is a good idea as well. Because there is not a design view in which you can visually lay out the different items, you will have to install your application in order to visually see your preferences' interface.

There are many different types of elements that can make up your application preferences' user interface, and they are explored throughout the course of this chapter. The first element you will most likely want to add is a Group type. The Group type is a way for you to implement groups of settings on a single interface page. It does not specify an application preference and contains only a string that is displayed before a user preference on-screen.

For example, you may have preferences for login information, sound effects, and a general group. Grouping similar types of preferences allows the users to quickly find the preference that they are searching for. This is especially important if you have a lot of preferences on the same screen.

To add a group to your preference settings, add a Dictionary item to the PreferenceSpecifiers array. A group Dictionary item contains two String entries, Type and Title. The Type key represents the type of interface element that will be shown in the Settings dialog box on the device. In this case, the value should be set to PSGroupSpecifier. The Title key represents the String that will appear as the group header in the application preferences' interface. You can specify any String that you want here. Any preferences that are added after this group will appear in the same group.

Add a Settings Group

Note: For more details on how to create a Settings.bundle file, see the preceding section, "Create a Settings Bundle."

Note: These steps work only on Mac OS X. If you are using a Windows machine, see the Apply It on the next page.

1. In Xcode, open the Root.plist file.
2. Click PreferenceSpecifiers.
3. Click the Add Item button.

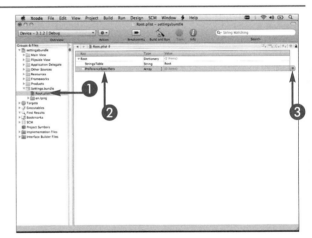

4. Click here and select Dictionary.
5. Select the row.
6. Click here to expand the item.
7. Click the Add Item button.

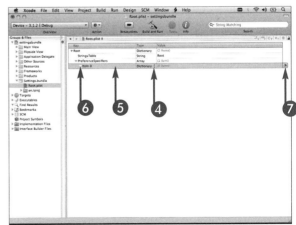

- ⑧ Click here and select String.
- ⑨ Name the key Type.
- ⑩ Give it a value of PSGroupSpecifier.
- ⑪ Select the row.
- ⑫ Click the Add Item button.

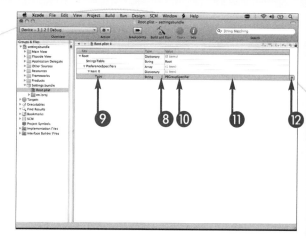

- ⑬ Click here and select String.
- ⑭ Name the key Title.
- ⑮ Give it a value, such as Group Header.

  A group is added to your settings bundle.

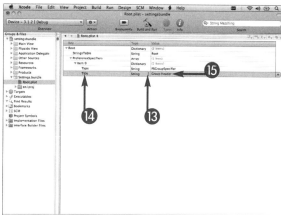

Apply It

The example in this section explores adding a Group header to the Settings file using the Property List Editor application on Mac OS X. If you are developing your application on a Windows machine, you can add the following text into your Root.plist file as a child of the `<array>` node:

```
<dict>
            <key>Type</key>
            <string>PSGroupSpecifier</string>
            <key>Title</key>
            <string>Group Header</string>
</dict>
```

Add a Text Setting

You can use a Text Field element to allow the user to input text as the value for a specific preference for your application. A Text Field element visually consists of two parts, a label and a text input field. To add a Text Field element to your preference settings, add a `Dictionary` item to the `PreferenceSpecifiers` array. The `Type` key is specified as a `String` and is a child of the `Dictionary` item that you created. Setting the value for the `Type` key to `PSTextFieldSpecifier` will set the element to be a Text Field element.

There are a number of different key value pairs that can be used with a Text Field element. The `Key` key is required. This key is a `String` type and is used to retrieve the value of the preference at runtime in code.

The `Title` key represents the string that will be displayed to the left of the text input field. The label for the title is left-aligned and will appear in boldface. If you omit this key, the text input field will expand to fill the width of the entire row.

The `DefaultValue` key can be used to specify a default value for the preference. If this key is not present, an empty string is returned.

If the entered text needs to be secure, such as a password field, you can include the `IsSecure` key and set it to `YES`. Setting this key will mask the input as dots in the text input field. If the key is omitted, the default is set to `NO`.

The `AutocapitalizationType` key specifies the autocapitalization style while the user is typing. The value must be one of the following values: `None`, `Sentences`, `Words`, or `AllCharacters`.

Add a Text Setting

Note: *For more details on how to create a Settings.bundle file, see the section "Create a Settings Bundle."*

Note: *These steps work only on Mac OS X. If you are using Windows, see the Apply It on the next page.*

1. Open the Root.plist file.
2. Click `PreferenceSpecifiers`.
3. Click the Add Item button.

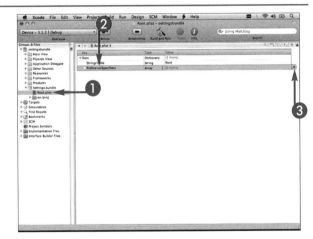

4. Click here and select `Dictionary`.
5. Select the row.
6. Click here to expand the item.
7. Click the Add Item button.

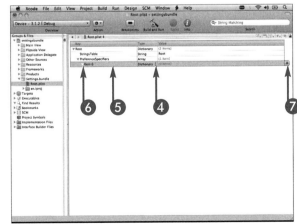

8 Add a Type key.

9 Set the value to PSTextFieldSpecifier.

10 Add a Title key.

11 Set the value, such as Username.

12 Add a Key key.

13 Set the value, such as username.

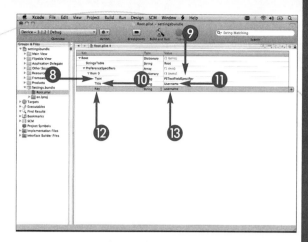

14 Add an AutocorrectionType key.

15 Set the value, such as No.

16 Add an AutocapitalizationType key.

17 Set the value, such as None.

An input text field is added to the settings bundle.

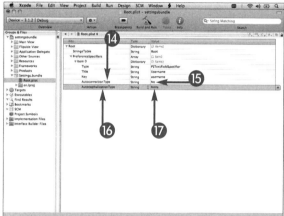

Apply It

The following is an example of the XML output for a Text Field element. You can use this as a template if you are developing on a Windows machine:

```
<dict>
            <key>Type</key>
            <string>PSTextFieldSpecifier</string>
            <key>Title</key>
            <string>Username</string>
            <key>Key</key>
            <string>username</string>
            <key>AutocorrectionType</key>
            <string>No</string>
            <key>AutocapitalizationType</key>
            <string>None</string>
</dict>
```

Add a Multiple Value Settings Field

Some preferences require the user to select from a list of available options, similar to a drop-down box on an HTML page. When the user selects the multiple value preference in order to change it, the Settings application displays a new page with all the possible values to choose from. The user is returned to the previous page after a value has been selected.

The value of the Type key for a Multiple Value setting is PSMultiValueSpecifier. The initial state of the preference looks similar to a Text Field element, with a title on the left and the value to the right. The Title key is required — the title that represents the preference. The Key key is a String value, which represents the preference key in which to associate the value. You can use the value of this key in order to retrieve the preference at runtime.

To specify the list of values that can be selected, you will add Values and Titles Array nodes to the main preferences Dictionary. The Titles key specifies a list of Strings, which represent the items in the Values array. The Titles are what is shown to the users when they are asked to make a selection. The Values is what the preference gets set to when the user makes a selection. Both keys are required, and they must have the same number of entries.

The DefaultValue key represents the default value for the preference. The value for this key must equal one of the entries in the Values key array.

Add a Multiple Value Settings Field

Note: *These steps work only on Mac OS X. If you are using Windows, see the Apply It on the next page.*

1. Add a Dictionary row to the PreferenceSpecifiers array.
2. Click here to expand the item.
3. Click the Add Item button.

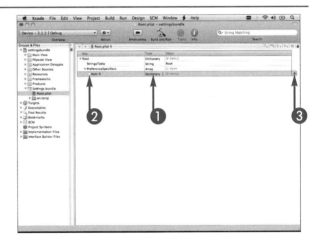

4. Add a Type key.
5. Set the value to PSMultiValueSpecifier.
6. Add a Title key.
7. Set the value, such as Email Type.
8. Add a Key key.
9. Set the value, such as emailtype.

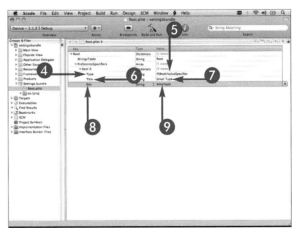

⑩ Add a `Values` key.

⑪ Select `Array` from the Type list.

⑫ Add a `String` item and set the value, such as `POP3`.

⑬ Add a `String` item and set the value, such as `SMTP`.

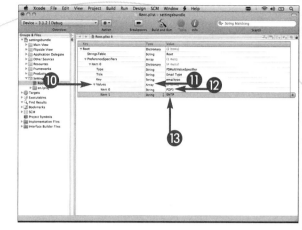

⑭ Add a `Titles` key.

⑮ Select `Array` from the Type list.

⑯ Add a `String` item and set the value, such as `POP3`.

⑰ Add a `String` item and set the value, such as `SMTP`.

⑱ Add a `DefaultValue` key.

⑲ Set the value, such as `POP3`.

A Multiple Value element is added to the settings bundle.

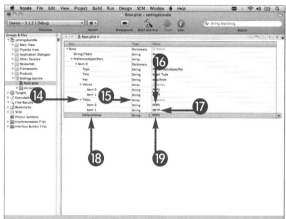

Apply It

The following is an example of the XML output for a Multiple Value element. You can use this as a template if you are developing on a Windows machine:

```
<dict>
                <key>Type</key>
                <string>PSMultiValueSpecifier</string>
                <key>Title</key>
                <string>Email Type</string>
                <key>Key</key>
                <string>emailtype</string>
                <key>Titles</key>
                <array>
                   <string>POP3</string>
                   <string>SMTP</string>
                </array>
                <key>Values</key>
                <array>
                   <string>pop3</string>
                   <string>smtp</string>
                </array>
                <key>DefaultValue</key>
                <string>pop3</string>
</dict>
```

Add a Toggle Switch Field

If there is a preference that can have only two values, you can implement a Toggle Switch element in your application preferences. This interface item is similar to the toggle that switches your device into Airplane mode, which is the very first preference in the Settings application.

To specify a Toggle Switch interface element, set the value for the `Type` key to `PSToggleSwitchSpecifier`. The item must implement the `Title` key, and you should set its value to a `String`, which will appear to the left of the toggle. The value for the `Key` key identifies the preference setting. Use the value for this key when attempting to retrieve the setting for the preference at runtime. The `Type`, `Title`, and `Key` are all required keys for the Toggle Switch element.

Visually, the Toggle Switch element shows "ON" and "OFF" as its labels. However, the values for each position are not limited to `Boolean` values. The `TrueValue` key represents the value of the preference when the toggle is set to the ON position. If this key is not present, the default value is a `Boolean` type with the value of `true`. The `FalseValue` key represents the value of the preference when the toggle is set to the OFF position. If this key is not present, the default value is a `Boolean` type with the value of `false`. Both of these keys can be any scalar type such as `Boolean`, `Date`, `Data`, `String`, or `Number`.

The `DefaultValue` key is required, and its value should be set to one of the specified values if they are set.

Add a Toggle Switch Field

Note: *These steps work only on Mac OS X. If you are using Windows, see the Apply It on the next page.*

① Click `PreferenceSpecifiers`.

② Click here to expand the row.

③ Click the Add Item button.

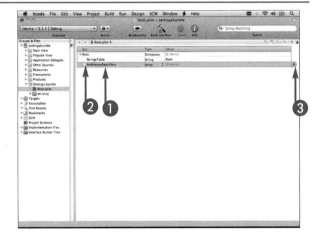

④ Click here and select `Dictionary`.

⑤ Select the row.

⑥ Click here to expand the item.

⑦ Click the Add Item button.

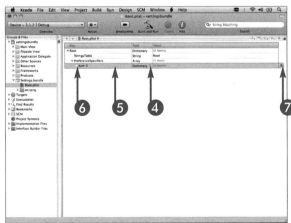

⑧ Add a Type key.

⑨ Set the value to PSToggleSwitchSpecifier.

⑩ Add a Title key.

⑪ Set the value, such as Sounds.

⑫ Add a Key key.

⑬ Set the value, such as sounds.

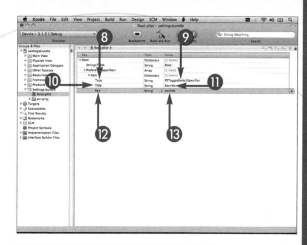

⑭ Add a FalseValue key.

⑮ Select Boolean from the Type list.

⑯ Add a TrueValue key.

⑰ Click here and select Boolean.

⑱ Check the box in the Value column.

A Toggle Switch element is added to the settings bundle.

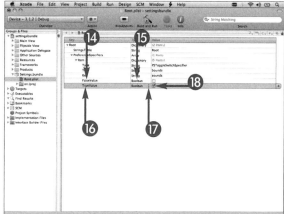

Apply It

The following is an example of the XML output for a Toggle Switch element. You can use this as a template if you are developing on a Windows machine:

```xml
<dict>
             <key>Type</key>
             <string>PSToggleSwitchSpecifier</string>
             <key>Title</key>
             <string>Sounds</string>
             <key>Key</key>
             <string>sounds</string>
             <key>TrueValue</key>
             <true/>
             <key>FalseValue</key>
             <false/>
             <key>DefaultValue</key>
             <true/>
</dict>
```

Add a Slider Settings Field

The Slider interface element can be used to allow the user to specify a value within a given range of values. Setting the `Type` key to `PSSliderSpecifier` will set the node to be a Slider interface element.

The Slider interface element does not have a `Title` key associated with it. Because of this, it is a good idea to place a group element before all sliders that will let the user know what the value represents. For more details on creating a Group header, see the section "Add a Settings Group" earlier in this chapter.

To specify the range of values to use in your slider, set the `MinimumValue` and `MaximumValue` keys. The value for these keys should be set to a `Number`. You will also need to set the `DefaultValue` key as the initial setting of the preference. You will need to make sure that the default value is within the range of the minimum and maximum values. All three of these keys are required in order for your slider element to work correctly.

The required `Key` key value is a `String` value, which is used to retrieve the value of the preference at runtime from within your application.

Because there is no `Title` key, you can also specify to have an icon appear at either end of the slider. The `MinimumValueImage` key represents the path to an image in the root of your Settings.bundle directory that will appear to the left of the slider. The `MaximumValueImage` key represents the path to an image in the root of your Settings.bundle directory that will appear to the right of the slider. The dimensions for each of these images should be 21 x 21 pixels.

Add a Slider Settings Field

Note: *These steps work only on Mac OS X. If you are using Windows, see the Apply It on the next page.*

① Create a `Dictionary` item in the `PreferenceSpecifiers` array.

② Create a `Type` key.

③ Type **PSGroupSpecifier** as the value.

④ Create a `Title` key.

⑤ Enter a value, such as `Sound Volume`.

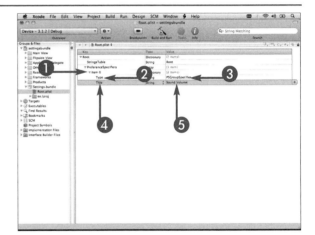

⑥ Create a `Dictionary` item in the `PreferenceSpecifiers` array.

⑦ Create a `Type` key.

⑧ Type **PSSliderSpecifier** as the value.

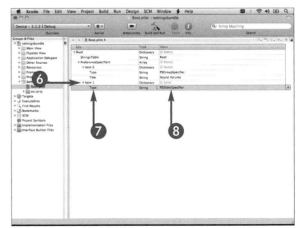

9 Create a Key key.

10 Enter a value, such as volume.

11 Create a DefaultValue key.

12 Select Number from the Type list.

13 Enter a value, such as 0.5.

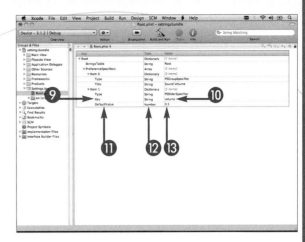

14 Create a MinimumValue key.

15 Select Number from the Type list.

16 Enter a value, such as 0.

17 Create a MaximumValue key.

18 Select Number from the Type list.

19 Enter a value, such as 1.

A slider is added to the settings bundle.

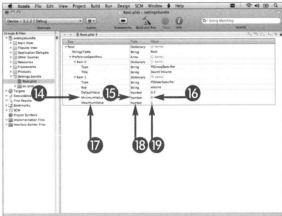

Apply It

The following XML is an example of a Slider element node in your Root.plist file. Adding this node to the PreferenceSpecifiers array will place a Slider element on the preference screen for your application. You can use this if you are developing on a Windows machine:

```
<dict>
            <key>Type</key>
            <string>PSSliderSpecifier</string>
            <key>Key</key>
            <string>volume</string>
            <key>DefaultValue</key>
            <real>0.5</real>
            <key>MinimumValue</key>
            <real>0</real>
            <key>MaximumValue</key>
            <real>1</real>
</dict>
```

Add a Title Settings Field

There may be instances when you want to display a read-only preference or property to the user. Many applications have used the Settings preferences to display the current version of the application. This gives the users an easy way to determine which version of the application they have installed without having to launch the application or having a spot in the application to view it.

To create a Title interface element, set the `Type` key value to `PSTitleValueSpecifier` as a `String`. The value of the `Title` key can be used as a human-readable title for the preference. This string is displayed to the left of the value. The `Key` key value is the identifying value, which is used to read the preference at runtime with code. This key is required; however, because this preference is read only, there is a good chance you will not need to read this preference in your application. Instead, you can simply create a static constant in your code to represent this value.

The `DefaultValue` key is used to set the value of the key. This key is required, and the value type is a `String`.

You can also use the Title element to inform your users about functionality. For example, if your application requires that you log in, it is a good idea to store the user's credentials as application settings. However, you may want to store only the username and not the password. Instead of leaving the password field out, you could put a Title element in that states that the password must be entered in the application.

Add a Title Settings Field

Note: *These steps work only on Mac OS X. If you are using Windows, see the Apply It on the next page.*

① Add an item to the `PreferenceSpecifiers` array.

② Click here and select `Dictionary`.

③ Click the Add Item button.

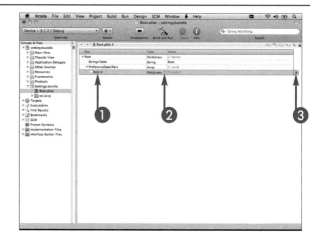

④ Create a `Type` key.

⑤ Set the value to `PSTitleValueSpecifier`.

⑥ Create a `Title` key.

⑦ Set the value, such as `Version`.

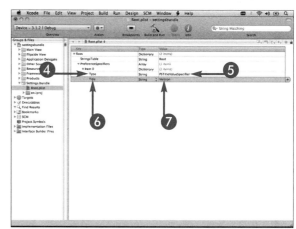

⑧ Create a Key key.

⑨ Set the value, such as version.

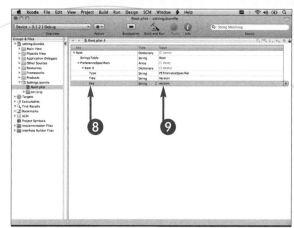

⑩ Create a DefaultValue key.

⑪ Set the value, such as 3.100.1.

A Title element is added to the settings bundle.

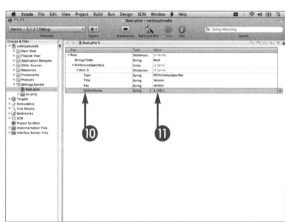

Apply It

The following XML is an example of a Title element node in your Root.plist file. Adding this node to the PreferenceSpecifiers array will place a Title element on the preferences screen for your application. You can use this if you are developing on a Windows machine:

```
<dict>
            <key>Type</key>
            <string>PSTitleValueSpecifier</string>
            <key>Key</key>
            <string>version</string>
            <key>Title</key>
            <string>Version</string>
            <key>DefaultValue</key>
            <string>3.100.1</string>
</dict>
```

Add the Settings Bundle to Your Application

After your Settings.bundle file is completed, you can copy the file from your temporary Xcode project and save it to a location in your current project directory structure. After the file is copied, you can feel free to delete the Xcode project and all associated files if you want to. If you need to make changes to the file, you can open the Root.plist file with the Property List Editor application that comes with Mac OS X. In order to gain access to the Root.plist file, right-click the Settings.bundle file in a Finder window and click Show Package Contents.

You can add your Settings.bundle file from the General tab of the iPhone Settings dialog box. Including the file will be different depending on which system you are developing on. If you are working on a Mac OS X machine, you can browse to the file by selecting the + button above the Included Files list. However, if you are developing on a Windows machine, you will need to select the folder icon to be able to select the Settings.bundle folder. Selecting just the folder will also bundle all its contents when your application is compiled. The Settings.bundle file must be added to the root directory of your application.

If you want to make sure that your Settings.bundle file was properly included in your application, you can change the extension of your .ipa file to .zip. Upon extracting the .zip file, you will find a Payload folder. Inside that folder is the .app version of your application. Inside there should be your Settings.bundle file.

Add the Settings Bundle to Your Application

1. Copy the Settings.bundle file to your Flash project directory.
2. In Flash, click File.
3. Click iPhone OS Settings.

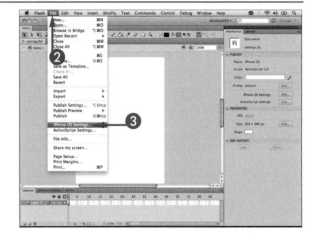

The iPhone Settings dialog box appears.

4. Click the General tab.
5. Click the + button.

Note: *If you are running Windows, click the folder button instead of the + button.*

274

The Open dialog box appears.

6 Navigate to and click your Settings.bundle file.

7 Click Open.

You are returned to the iPhone Settings dialog box.

• The Settings.bundle file is now bundled with the application.

8 Click OK.

9 Compile and install the application and open the Settings application on the device.

Your application appears toward the end of the main settings screen.

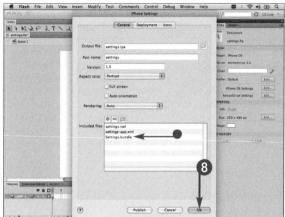

Extra

When planning your preferences file, you should think about how regularly the users will be updating the preferences. If the preferences will be updated on a regular basis, it may be a good idea to implement the preferences within the application instead of the Settings application. If the only place the users can update the preferences is in the Settings application, they will have to exit your application to edit them.

Also, if your application has only a few preferences, it might make more sense to keep them in the application. Integrating the preferences in the Settings application is great if you have many preferences and need a lot of real estate to display them all. This is especially useful for Flash iPhone applications because they currently do not support the native iPhone SDK components, so you would have to develop them on your own.

As a general rule, you should not use both methods to display your preferences. However, there are use cases where it is acceptable to break this rule — for example, login credentials. You will need to log in from within the application, but it is also convenient to be able to change the credentials without having to launch the application.

Read the Settings

After you have created the Settings.bundle file and have added it to the list of files to be included with your application, you are ready to write some code in order to retrieve the preference values. It is important to understand that the Settings.bundle file simply configures the user interface of your preference screens. When a preference is updated, the value is not reflected in the Settings.bundle file.

A binary property list file is created in the application file sandbox, which has key value pairs for each of the preferences in your Settings.bundle file. The file is stored in the <Application Home>/Library/Preferences folder and has a filename the same as your application ID. You can use the following syntax to read the file:

```
var path:String ="Library/Preferences/"+
  NativeApplication.nativeApplication.
  applicationID+".plist";
var file:File = File.userDirectory.
  resolvePath( path );
```

For more information on working with local files, see Chapter 11, "Working with Files." Because the file is not in a text-based format, you will have to understand the binary property list file format in order to parse the key value pairs out of the file. To help with this, there is an `IPhoneAppSettings` class provided with the source of this book that will parse the file for you.

The `IPhoneAppSettings` class is a singleton class, which when instantiated for the first time will load the preferences file. The file is opened synchronously, which means that you can try to retrieve preference values immediately after you have instantiated it.

Read the Settings

Note: In order to read the settings, you must first have created a Settings.bundle file and bundled it with your application. See earlier sections in this chapter for more information.

① Click the Text tool.

② Create a text field on the Stage.

③ Click here and select Classic Text.

④ Click here and select Dynamic Text.

⑤ Give the text field an instance name, such as `settings_txt`.

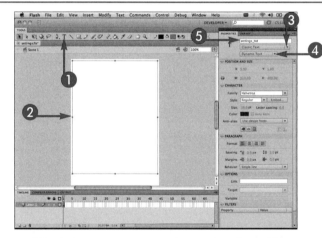

⑥ Click the New Layer button.

A new layer is created.

⑦ Select a layer to which to add ActionScript code.

⑧ Open the Actions panel.

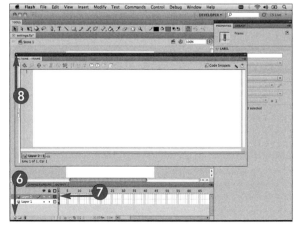

9. Create an `IPhoneAppSettings` variable, such as `var settings:IPhoneAppSettings`.

10. Get a reference to the `IPhoneAppSettings` instance, such as `IPhoneAppSettings.getInstance();`.

11. Append text to the text field, such as `settings_txt.appendText();`.

12. Append the version key, such as `settings.stringForKey( "version" )`.

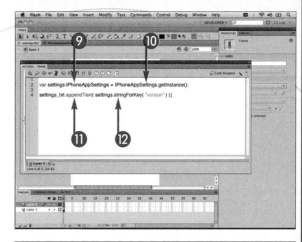

13. Append text to the text field, such as `settings_txt.appendText();`.

14. Append the version key, such as `String( settings.boolForKey( "sound" ) )`.

15. Append text to the text field, such as `settings_txt.appendText();`.

16. Append the version key, such as `String( settings.numberForKey( "volume" ) )`.

The values for the version, sound, and volume settings are placed in the text field on the Stage.

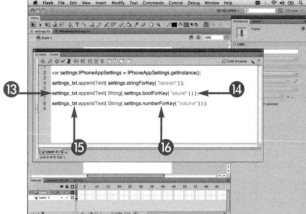

Extra

The code for this class can also be found in the open source as3iphone project at http://code.google.com/p/as3iphone/. It is a good idea to make sure that you always have the most up-to-date code, which may include any bug fixes or new features.

As well as read the preferences, you can also set new values for a preference. In most cases, you will want to have the preferences set only in the Settings application. However, in case you want to save them with code, you can call the appropriate setting method on the `IPhoneAppSettings` class. After a preference has been set, all the preferences are re-encoded to the binary property list format and saved to the local file system of the device. The following syntax is an example of how to set a new value for the `username` key:

```
bplist.setStringForKey( "username", "thomas" );
```

Add an Icon to Your Settings

When the Settings application is launched, it checks each custom third-party application installed on the device and checks to see if a Settings.bundle file exists. For each Settings.bundle file that it finds, it loads the bundle and displays the icon and application name it is associated with.

You can specify a custom icon to represent your application in the Settings application third-party application list. The image dimensions for the icon should be set to 29 pixels wide by 29 pixels high. Naming your icon Icon-Settings.png and placing it in the root directory of your application will cause it to be displayed in the Settings application list.

If the custom icon file cannot be found, the Settings application will display the default icon for your application, Icon.png. It will also perform any necessary scaling on the icon in order to make sure that it fits correctly in the row to the left of the application name. Scaling down the image for the application icon can cause the icon to appear pixilated or distorted. It is highly recommended that you create a smaller version of your icon image and place it in the root of your Settings.bundle folder. This will ensure that your icon appears crisp and maintains its design integrity.

Your custom icon should simply be a smaller representation of the icon that is used on the home screen of the device. Creating a different icon can be confusing to your users. Keep this in mind when designing your application icon. An icon with a lot of detail may not translate very well when scaled to smaller sizes.

Add an Icon to Your Settings

1. Open an image-editing application, such as Photoshop.
2. Open your application icon file.
3. Click Image.
4. Click Image Size.

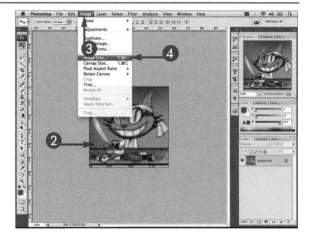

The Image Size dialog box appears.

5. Set the width to 29 pixels.
6. Set the height to 29 pixels.
7. Click OK.
8. Click File → Save As.
9. In the Save As dialog box, save the image as Icon-Settings.png.

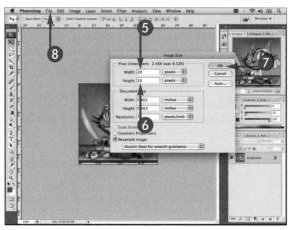

⑩ In Flash, click File → iPhone OS Settings.

The iPhone Settings dialog box appears.

⑪ Click the + button.

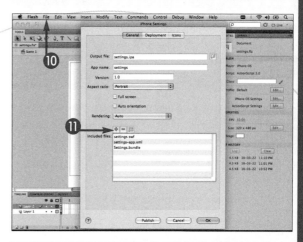

The Open dialog box appears.

⑫ Navigate to and click your Icon-Settings.png file.

⑬ Click Open.

You are returned to the iPhone Settings dialog box.

⑭ Click OK.

When you compile and install the application on your device, you will see the icon appear in the Settings list.

Apply It

If you created your Settings.bundle file in Xcode, a localization folder is generated for you, such as en.lproj. Inside this folder is a file named Root.strings, which contains a list of strings that can be localized in your application settings file. Any `Title` keys in any of the nodes can be localized. If the string is found in the strings file, the Settings application will insert the text based on the language that the device is set to. If it does not find the string, the Settings application will use the value for the `Title` key. Here is a sample of the Root.strings file:

```
"Sound" = "Sound";
"Volume" = "Sound Volume";
"Username" = "Username";
"Password" = "Password";
```

Read Your Device's Global Settings

Along with your application settings, you can read some of the global preferences for the device. There is a hidden file in the <Application Home>/Library/Preferences named .GlobalPreferences.plist. This file is a binary property list file that has some very useful preferences in it.

Included with the source of this book is a `GlobalSettings` class. The `GlobalSettings` class is a singleton class that reads and parses the .GlobalPreferences.plist file. Because the .GlobalPreferences.plist file is created by iOS, new values should not be added to it. Also, you can easily find out all the key value pairs that are included in the file. The `GlobalSettings` class has a helper method to return each of the preferences included in the file.

The `GlobalSettings.get24HourClock()` method returns `true` if the device is set to display the current time in 24-hour mode and `false` if it uses the 12-hour mode.

The `GlobalSettings.getAppleKeyboards()` method returns an `Array` of `String`s that represent all the available keyboard locales on the device. This array also represents all the keyboards that you have turned on in the Settings → General → International → Keyboards list.

The `GlobalSettings.getAppleLanguages()` method returns an `Array` of `String`s that represent all the available languages on the device. This is the representation of all the languages in the Settings → General → International → Language list.

The `GlobalSettings.getPhoneNumber()` method returns a `String` of the current phone number of the device. This preference is available only when the Phone application is installed on the device.

Read Your Device's Global Settings

1. Click the Text tool.
2. Create a text field on the Stage.
3. Click here and select Classic Text.
4. Click here and select Dynamic Text.
5. Give the text field an instance name, such as `settings_txt`.

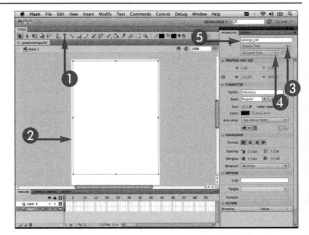

6. In the Actions panel, create a `GlobalSettings` reference, such as `var settings:GlobalSettings=GlobalSettings.getInstance();`.
7. Append the text for the 24-hour clock, such as `settings_txt.appendText( "24 hour clock = " + settings.get24HourClock().toString() + "\n" );`.

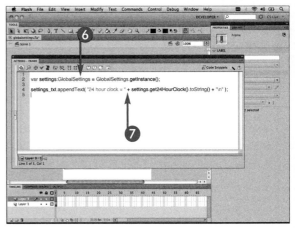

⑧ Append the text for the keyboards, such as
```
settings_txt.appendText(
"keyboards = " + settings.
getAppleKeyboards().toString() +
"\n");.
```

⑨ Append the text for the languages, such as
```
settings_txt.appendText(
"languages = " + settings.
getAppleLanguages().toString() +
"\n");.
```

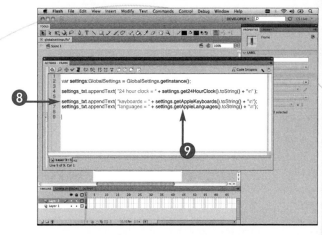

⑩ Append the text for the phone number, such as
```
settings_txt.appendText( "phone
number = " + settings.
getPhoneNumber().toString() +
"\n");.
```

The values for the 24-hour clock, keyboard locale, language, and phone number settings are placed in the text field on the Stage.

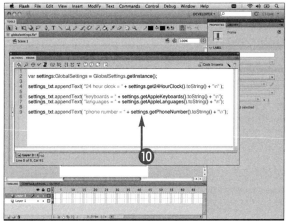

Extra

There are a few other preferences in the .GlobalPreferences.plist file that were not explored in the example here. The following method will return a String value that represents the current locale of the device. This string is usually a combination between the current language and the country:

```
GlobalSettings.getInstance().getLocale();
```

The final two methods in the .GlobalPreferences.plist file are `GlobalSettings.getTVOutStatus()` and `GlobalSettings.getAppleKeyboardExpanded()`. Both methods return an int.

Show Your Trace Statements

If you have been developing Flash applications for quite some time, you are probably very familiar with the `trace()` method. The `trace()` method enables you to output values to the Output panel within the Flash IDE. You can bring up the Output panel by pressing the F2 keyboard shortcut in the IDE.

The `trace()` statement can have multiple parameters passed into it. If any of the arguments are not of a `String` data type, the `trace()` method invokes the `toString()` method on that object. All arguments are outputted to the Output window on a single line with multiple arguments separated by a space.

The remote debugging capabilities of the Flash IDE allows you to receive your `trace` statements in the IDE when testing on the device. In order for this to work, your device must have WiFi turned on and be connected to the same network as the computer running the Flash IDE.

You will also need to set the iPhone deployment type to Quick Publishing for Device Debugging in the iPhone Settings dialog box. When your application is published in this mode, you will be asked to enter the IP address of your computer when your application is launched.

If the Flash IDE is waiting for a remote player to connect, the application will attempt to connect to your computer over WiFi and start a remote debugging session. The option Begin Remote Debugging Session on the Debug menu of the Flash IDE enables you to set up the IDE to receive the connection from your application. This should be done before you enter your IP address on the device.

The IP address of your computer can be found in the Network Preferences on Mac OS X or by running the `ipconfig` command from a command line in Windows.

Show Your Trace Statements

① Open the Actions panel.

② Create some `trace` statements.

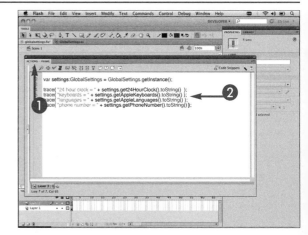

③ Click File → iPhone OS Settings.

The iPhone Settings dialog box appears.

④ Click the Deployment tab.

⑤ Click Quick Publishing for Device Debugging.

⑥ Click OK.

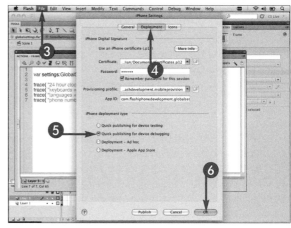

7. Click Debug.

8. Click Begin Remote Debug Session.

9. Click ActionScript 3.0.

10. Compile and install the application on your device.

11. When prompted, enter the IP address of your computer.

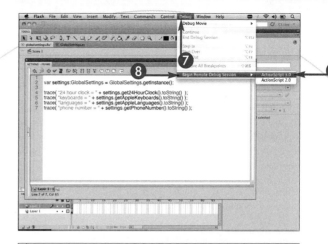

- The workspace is changed to DEBUG.
- The `trace` statements are shown in the Output panel.

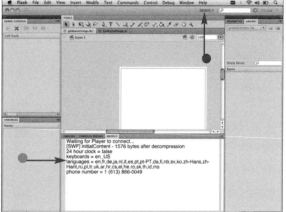

Chapter 17: Debugging Your Application

Apply It

The Output panel and the `trace()` method work well for outputting very simple expressions. They do not do well, however, when dealing with complex objects. There are several alternatives available to you. The POWERFLASHER SOSmax Socket Output Server is a great tool for more complex output from Flash. It runs as an XML-based socket server on your computer, which you can connect to from your device. Here is an example of how to connect to the application from your device:

```
var socket:XMLSocket = new XMLSocket();
socket.addEventListener( Event.CONNECT, onConnect );
socket.connect( "10.0.0.1", 4444 );
function onConnect( e:Event ):void{
socket.send( "connected to device" );
}
```

283

Create Breakpoints

When you are trying to debug a complex issue or problem with your code, the `trace()` method may not give you enough information. If there is a lot of code executing at a given time and it is outputting lots of `trace` statements to the Output panel, it may be difficult to decipher where the problem is. Having the ability to pause your application from continuing allows you to at any given time see the result of a line of code.

Setting a breakpoint lets you stop the application at a specific line of code. It is important to understand that there must be valid code on the same line as the breakpoint. For example, if you set a breakpoint on a line without code or a comment, the application will not stop at that breakpoint.

Breakpoints can be set in the Actions panel, the Script window, and the Debugger. All breakpoints set in the Actions panel are saved with the FLA and will be present the next time you open the file. Breakpoints set in the Script window and Debugger are not saved and are available only for the current development session.

When editing an external ActionScript file in the Script window, there are several shortcut commands that can be used to add and remove breakpoints. Pressing the ⌘+B (Ctrl+B) keyboard shortcut will toggle a breakpoint at the line of code on which the cursor has focus. There are also options from the Debug main menu item that allow you to remove all the breakpoints for a given file or from all ActionScript files.

Create Breakpoints

Set a Breakpoint in the Actions Panel

1. Open the Actions panel.
2. Click in the left column.

   A breakpoint is added.

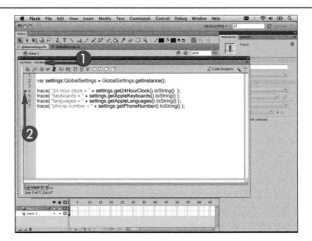

Remove All Breakpoints in the Actions Panel

1. Right-click in the window.
2. Click Remove Breakpoints in This File.

   All the breakpoints in the FLA are removed.

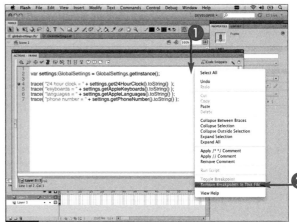

Set a Breakpoint in the Script Window

1. Click File → Open.
2. Navigate to and select an ActionScript 3.0 class file.
3. Click Open.
   - The class file opens in the Script window.
4. Click in the left column.
   A breakpoint is added.

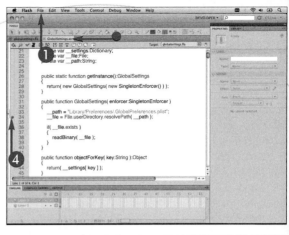

Remove All Breakpoints in the Script Window

1. Click Debug.
2. Click Remove Breakpoints in This File.
   All the breakpoints in the class are removed.

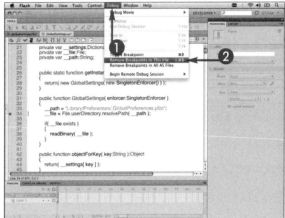

Extra

Breakpoints added in the Script window can also be saved in the AsBreakpoints.xml file, which is located in the configuration directory for Flash.

On a Windows XP machine, the AsBreakpoints.xml file can be found at the following location:

Windows Hard Drive\Documents and Settings\User\Local Settings\Application Data\Adobe\Flash CS5\language\Configuration\Debugger\

On a Windows 7 machine, the AsBreakpoints.xml file can be found at this location:

Windows Hard Drive\Users\<username>\AppData\Local\Adobe\Flash CS5\en_US\Configuration\Debugger

On a Mac OS X machine, the AsBreakpoints.xml file can be found at the following location:

Macintosh HD/Users/User/Library/Application Support/Adobe Flash CS3/Configuration/Debugger/

AsBreakpoints.xml is a simple XML file that has a node for each file containing a breakpoint. For each breakpoint in the file, a child node is created with the line in which the breakpoint exists as an attribute.

Using the Flash CS5 Debugger

After you have set some breakpoints in your code, you can use the Flash CS5 ActionScript 3.0 Debugger to have your application pause at each breakpoint. Debugging your application works the same for any Flash application, whether it is running on a device or running locally on your computer.

In some instances, you will need to specify to permit debugging in the Flash Publish Settings dialog box; however, specifically for iPhone development, you can omit this step. Debugging your application is different from testing your movie, or pressing ⌘+Enter (Ctrl+Enter). When you open the Debugger, selecting the appropriate debug scenario, your application will be compiled, and the Flash IDE will switch to the Debug workspace.

The Debug workspace displays panels that you can use to help debug your application effectively. You can also select this workspace from the workspace drop-down list located at the top right of the application bar.

The Debugger will pause at any line of code that throws an `Error`. Most of the time `Errors` are thrown because the Debugger has encountered a `null` variable. Having the Debugger stop on the specific line that caused the `Error` enables you to easily determine the `null` variable.

When you are using the Flash Debugger to debug a remote application, the Debugger will wait for two minutes for the remote application to connect. If a connection is not made within the two minutes, the Debugger times out, and you have to restart the remote debug session.

After a debug session has started, it can be stopped by clicking End Debug Session on the Debug main menu. You can also close the compiled file in order to end the session.

Using the Flash CS5 Debugger

1. Click File.
2. Click Publish Settings.

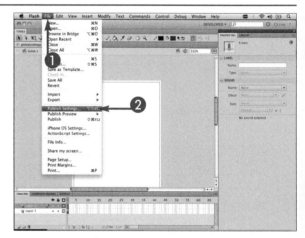

The Publish Settings dialog box appears.

3. Click the Flash tab.
4. Click Permit Debugging.
5. Click OK.

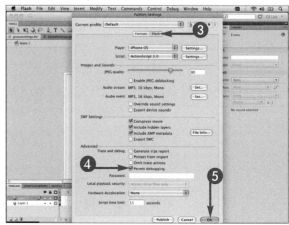

6 Click Debug.

7 Click Debug Movie.

8 Select where you would like to debug your application, such as In AIR Debug Launcher (Mobile).

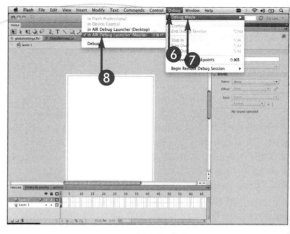

- The workspace is changed to DEBUG.
- The application is compiled and launched for debugging.
- The application will pause at any breakpoint.

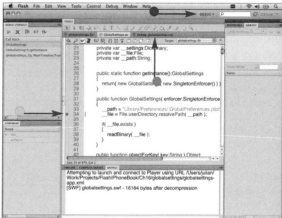

Chapter 17: Debugging Your Application

Extra

Even though this book is about creating Flash iPhone applications, there are many elements in the book that apply to any Flash application. This is especially true when debugging your application. If you are deploying your Flash applications to a Web server, there may be times where you will need to debug the application on the Web server through a Web browser. Because the compiled Flash file is not on your local file system, certain restrictions are in place for security reasons.

However, there are many browser plug-ins that you can install in order to receive the `trace` statements from your application. One of the most popular is a Firefox plug-in called FlashTracer. It provides you with a window that gathers all the `trace` statements from a Flash SWF file with Permit Debugging turned on. Running any of these types of browser plug-ins can have an effect on the performance of your application. When you have tracked down and eliminated the bugs, turn the plug-in off to see how your application performs.

Understanding the Debug Console

The Debug Console contains many useful features of the Flash CS5 Debugger. In most instances, this panel should automatically be shown when the Flash IDE workspace is changed to Debug. However, if for whatever reason it does not appear, you can show it by selecting it on the Debug Panels menu from the Window main menu.

There are two important parts of the Debug Console, the top menu bar and the Call Stack list. The top menu bar consists of a set of controls that allows you to control the Debugger. When the Debugger pauses at a breakpoint, you can press any of the buttons in the top bar to continue executing the application.

The Continue button continues running the application until it reaches the next breakpoint. The End Debug Session button closes the Debugger and the application, ending the current session. The Step Over button executes the line of code at which it has stopped and then pauses the application on the next line of code. The Step In button causes the Debugger to execute the current line of code and enter the Break mode. If the line of code calls another method, it will then step into that method. If you step into a different method, you can choose to step over or into each line, just as you would if the Debugger had stopped on a breakpoint. You can also press the Step Out button in order to return to the method that you step in from.

When the Debugger pauses your application from executing, the Call Stack shows a list of the remaining functions that are waiting to finish executing. Click an item in the list to jump to that method or script.

Understanding the Debug Console

① Add a breakpoint to your application.

② Start debugging your application.

Note: For more details on adding breakpoints and debugging your application, see the sections earlier in this chapter.

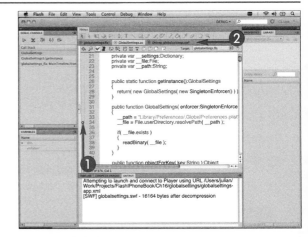

③ Click the Step Over button.

● The Debugger jumps to the next line of code to execute and pauses.

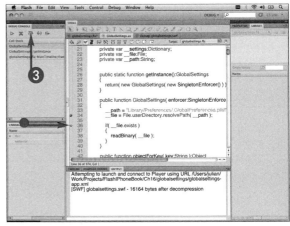

④ Click an item in the Call Stack list.

● The Debugger navigates to that location in the code.

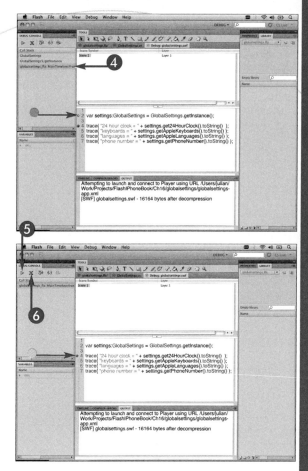

⑤ Click the Continue button.

● The Debugger continues executing code until it reaches another breakpoint.

⑥ Click the End Debug Session button when you are finished debugging your application.

The debugging session closes.

Extra

In order to make your debugging experience as efficient as possible, it is a really good idea to memorize the keyboard shortcuts for the stepping methods in the Debug Console. Being able to quickly jump in, out, and over code will increase the speed in which you can debug your application. You can find the keyboard shortcuts by selecting Debug from the main menu. The shortcut key is located to the right of the item.

If you find that the default keyboard shortcut keys are not convenient for you, you can change them to something that works better for you. To customize the keyboard shortcuts, click Flash ➔ Keyboard Shortcuts on a Mac OS X machine or Edit ➔ Keyboard Shortcuts on a Windows machine.

Before you can customize the keyboard shortcuts, you must first create a duplicate of the default keyboard shortcuts. To duplicate the set, click the Duplicate Set button, which is the first button to the right of the Current Set drop-down list. Each set is stored in an .mfx file in the Flash Configuration folder. This makes it easy to share shortcuts with other members of your team.

Understanding the Variables Panel

The Variables panel lists all the current objects and variables in the current scope. By default, the Variables panel should be shown when Flash switches to the Debug workspace. However, if for some reason it does not appear, you can show it using the Debug Panels menu on the Window main menu.

The Variables panel consists of two columns, Name and Value. The Name column shows the name of the variable in your code, and the Value column shows the value of that variable. This is extremely useful for tracking down any `null` variable `Error`s, as sometimes it can be hard to tell which object is `null` based on the `Error` message.

For any complex objects, such as classes, objects, arrays, and dictionaries, you can expand them to view their contents and properties. Being able to drill into objects within `Array`s and `Dictionary` objects is an extremely powerful tool in debugging your application. Without the Debugger, you would have to perform some heavily nested `if` statements in order to properly `trace` the objects and properties that you are interested in.

You can also change the value of a variable in the Variables panel. Double-clicking the Value column of the variable that you want to edit will change the row into an editable text input. After you edit the value, the new value will be used during the next code execution.

By default, there are certain types of variables that are hidden from the Variables panel. Clicking the panel preferences button will bring up a list of all available types to be shown.

Understanding the Variables Panel

① Add a breakpoint to your application.

② Start debugging your application.

Note: For more details on adding breakpoints and debugging your application, see the sections earlier in this chapter.

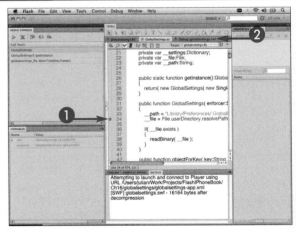

③ Click the arrow to expand any complex objects in the Variables panel.

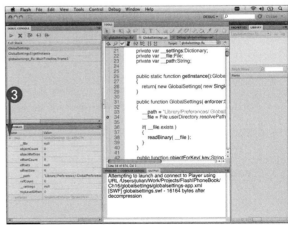

④ Click an item in the Call Stack list.

⑤ Click the arrow to expand any of its complex objects in the Variables panel.

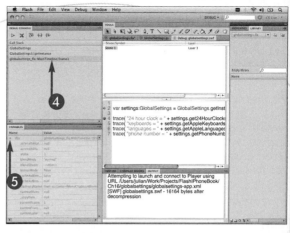

⑥ Double-click one of the values to change.

⑦ Edit the value.

⑧ Press Enter to exit edit mode.

The value of the property is changed in your application.

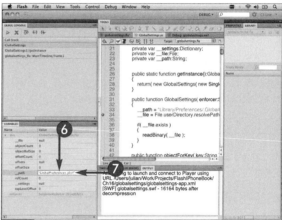

Chapter 17: Debugging Your Application

Extra

Changing variables on the fly can be used for more than just debugging. Being able to tweak parameters in a physics simulation or game can be useful, as it does not make you recompile the application in order to see the changes. It also enables you to test the effects on your application if variables are set to different values.

The Variables panel is updated when you click an item in the Call Stack list. This potentially allows you to track down bugs because you will be able to see the evolution of a variable as each item in the call stack is executed. It also allows you to see which variables are available within a given scope.

If you try to change a variable to a data type other than its current data type, the Variables panel will revert the value to its last valid value. For example, changing an int variable value to a String is not permitted through the Variables panel.

Unfortunately, you are not able to instantiate variables from the Variable panel. For example, an Array variable whose current value is null cannot be changed.

291

Get Crash Reports

Your application will crash at some point, so you need to get used to it. Even when you think you have found all the bugs, your application will crash sometime on someone's device. It is okay if your application crashes, and you should be positive about the situation and use the tools at your disposal in order to fix any bugs that may be causing the crash.

When a crash occurs, an iOS device saves the crash report to the device. These reports describe what caused the application to crash. When a user syncs his or her device using iTunes, crash reports stored on the device are copied to a directory on the machine. If the user has selected to submit crash reports to Apple, they will be uploaded to Apple, and you can download them from the iTunes Connect site.

The reports that are saved on the user's machine can be found in the following directories: on Mac OS X, in ~/Library/Logs/CrashReporter/MobileDevice/<DEVICE_NAME> and in Windows 7 and Vista, in C:\Users\<USERNAME>\AppData\Roaming\Apple Computer\Logs\CrashReporter\MobileDevice\<DEVICE_NAME>.

There are four different types of reports. The first is an application crash, which is the most common. This can be caused by bad access or another programming error. The second type is from low memory. iOS will kill the currently running application if the system fails to free up enough memory. So if your application uses too much memory, it can be killed. The third type occurs when the user presses the home button down for six seconds to force-quit the application. These crashes can be identified by the 0xdeadfa11 exception code. The final type of crash is a watchdog timeout crash. This type of crash can be identified with the exception code 0x8badf00d and occurs if the application takes too long to launch or terminate.

Get Crash Reports

Note: Xcode is available only on Mac OS X. In Windows, you can find crash reports in C:\Users\<USERNAME>\AppData\Roaming\Apple Computer\Logs\CrashReporter\MobileDevice\<DEVICE_NAME>.

1. Open Xcode, which can be found in the /Developers/Applications folder.
2. Click Window.
3. Click Organizer.

The Organizer window appears.

4. Connect your device and select it.
5. Click the Crash Logs tab.
6. Select a crash report from the list.

Details of the report appear in the lower pane.

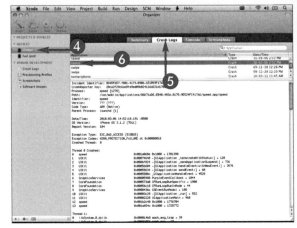

292

- The date and time of the crash appear here.
- This shows the OS version of the device, which caused the crashed.

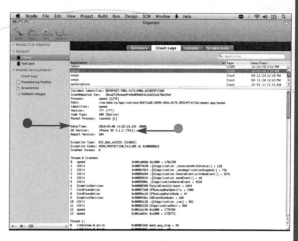

- Here is the exception type explaining why the application crashed.
- This specifies what thread crashed.
- This specifies what library caused the crash.

Chapter 17: Debugging Your Application

Extra

The iPhone and iPod touch have a very limited amount of memory available. Some versions have only 128MB of memory, whereas some of the newer versions have 256MB. This does not mean that your applications get to use all of this memory. A good portion of it is already being used and taken up by doing the everyday tasks of your user's device. For example, graphics takes up 12MB, the kernel takes up 32MB, daemons such as mail and mediaserverd take 12MB, Springboard takes 10MB, and the phone uses 4MB of memory. This instantly cuts down on the amount of available memory for your application. This does not even include the base overhead of a Flash iPhone application. It is really important that you manage your memory throughout the development of your application. Keeping a close eye on it will reduce the amount of low memory crash reports you receive. If you can measure the memory consumption of your application iteratively through development, it will be easier to isolate what portion of your application is taking up the most memory.

Using Instruments

Bundled with the Apple Developer tools is an application called *Instruments,* which is only available on Mac OS X. Instruments allows you to measure and track one or more processes in your application. Instruments is very similar to the Profiler included with Flash Builder and FDT4, but with more in-depth features. Instruments provides a series of templates to profile different portions of your application. These templates are known as *instruments.*

The Instruments application can be found in the Applications folder of your Developer Tools directory; by default, this is /Developers/Applications. With your iOS device connected to your computer through USB, you can select the application that you want to profile.

After your application has been selected, you have Instruments launch it on your device and begin recording any information that you have requested. You can set up more than one different type of instrument at any given time. Each time you run the application, the data is saved, allowing you to make changes and compare your results to hopefully see improvements. To compare the data between different runs, expand the instrument in the Instruments pane.

The Library panel lists all the available instruments that you can add. If you select iPhone from the drop-down list at the top of the panel, only the instruments that are available for the iPhone will be shown.

You can also add flags to the Track pane, which act as markers for specific moments in time. This allows you to mark when a specific event took place within your application so that you can revisit the data when you have finished profiling.

Using Instruments

Note: *You can perform these steps only on Mac OS X. If you are using Windows, try Flash Builder's Profiler for similar utility.*

1. Open Instruments.
2. Click All under the iPhone header.
3. Click a template, such as Blank.
4. Click Choose.

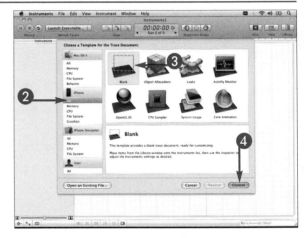

An empty document appears.

5. Click the Library button.

   The Library panel appears.
6. Click here and select iPhone.
7. Click an instrument, such as I/O Activity, and drag it to the Instruments pane.

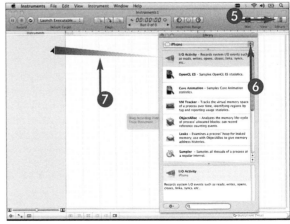

8 Click here.

9 Select your connected device.

10 Click Launch Executable.

11 Click an application installed on the device to profile.

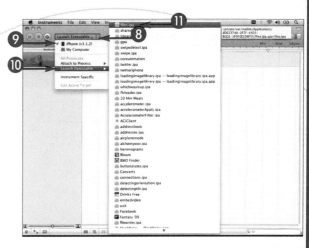

12 Click the Record button.

13 Click the Extended Detail button.

• Profiled data is graphed in the Track pane.

• Detailed data is shown in the Details pane.

Chapter 17: Debugging Your Application

Extra

Some of the instruments make use of the Extended Detail pane. Having it visible at all times will ensure that you do not miss any important data when profiling your application. To show the Extended Detail pane, click View ➔ Extended Detail or press the ⌘+E keyboard shortcut.

The Run browser enables you to quickly view all the runs for that trace document in the cover flow browser style. This allows you to quickly compare the data between all the runs. To view the Run browser, click View ➔ Run Browser. The Run browser also gives you the ability to enter comments for each run to better distinguish one from another.

Instruments is a very complicated and deep application. Apple has some great documentation and user guides on its developer portals that will help you better understand how to get the most out of your applications. Throughout the rest of this chapter, you will explore some of the different types of instruments and how you can use them to better understand how your application is performing on the device.

Using the ObjectAlloc Instrument

The Instruments application includes an instrument called *ObjectAlloc,* which is used for tracking all the memory allocation in your application. Note, again, that Instruments is available only on Mac OS X. Having the ability to see how your application is using memory enables you to tune specific portions of your application in order to try and reduce the memory as much as possible.

To start, select the ObjectAlloc instrument from the Library panel and drag it to the Instruments pane. In the Details pane, you can select to show All Objects Created, the objects that have been Created & Still Living, or the objects that have been Created & Destroyed. With your application selected in the Default Target drop-down list, you are ready to start recording your application.

When the application is launched, it will start recording data to the Track pane and the Details pane. The Track pane graphs the total memory consumption of your application over time. This enables you to quickly see any spikes in memory consumption during specific events throughout your application. It also shows which portions of your application are the most intensive on the memory of the device. Being able to isolate these areas allows you to fine-tune specific portions in order to reduce the memory of your application.

The Details pane breaks down each type of object that is currently using memory in your application. The Live Bytes column represents how much memory that specific object is currently using in the application. The # Living column represents how many of those objects are currently in memory. The Overall Bytes column shows how much memory over the course of the life of the application the object has used. The # Overall column shows how many instances of the object have been created during the life of the application.

Using the ObjectAlloc Instrument

Note: *You can perform these steps only on Mac OS X. If you are using Windows, try Flash Builder's Profiler for similar utility.*

1. Open Instruments.
2. Create an ObjectAlloc instrument.

Note: *See the preceding section, "Using Instruments," for more information.*

3. Click All Objects Created.
4. Click here and select an application to profile.
5. Click Record.

- Object allocation is graphed in the Track pane.

6. Click this column header to sort by Live Bytes to show the item consuming the most memory.
7. Click the arrow on an item to show a breakout of all instances.

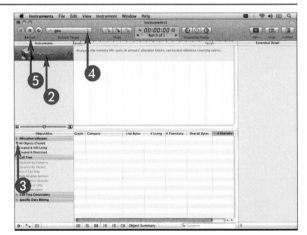

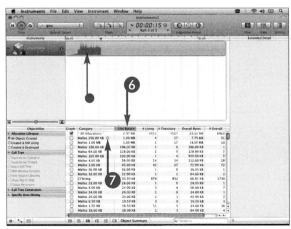

8. Click an instance in the list.
- The call stack is shown in the Extended Detail list.
9. Click the arrow button for more details for that memory address.

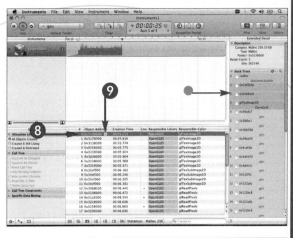

- A complete history of that memory address is shown in the Details pane.

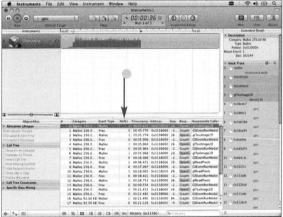

Extra

When you hover over a value in the Category column of the Details pane, a gray circle with a white arrow appears to the right of the text. Clicking this icon will show a list of all the instances created during the life of the application. The Details pane will change to a new grid of values. The Object Address column specifies the memory address for the allocated object. The Creation Time column shows at what time of the run the object was created. The Live column specifies if the object was still living or if it was destroyed. The Responsible Library column shows which Library was responsible for creating the object. This column can be extremely valuable in determining which portion of your application the object is being created from.

Because ObjectAlloc does not give you the names of the ActionScript classes, you will need to do some guesswork in determining which objects relate to the objects in your application. It is a good idea to isolate portions of your application in order to narrow down which objects could be created at a given time.

Using the Core Animation Instrument

The Instruments application includes an instrument called *Core Animation*, which can be used to profile animations. Note, again, that Instruments is available only on Mac OS X. The Track pane graphs the frames per second of any animations over the life of the application. It also shows the current frames per second in the far right of the pane.

When you are using Core Animation, the Track pane can also give you visual hints to help you better understand how graphics are being rendered on the screen. The Debug Options panel in the Details pane enables you to turn these hints on and off to see their effects.

Selecting the Color Blended Layers option puts a red overlay over layers that were drawn with blending enabled. It puts a green overlay over layers that were drawn without blending. The Color Copied Images options puts a cyan overlay over images that were copied by Core Animation. The Color Immediately option specifies to not wait 10 ms after performing a color-flush operation. The Color Misaligned Images option puts a magenta overlay over images whose source pixels are not aligned to the destination pixels. The Color Offscreen-Rendered Yellow option puts a yellow overlay over off-screen–rendered content. The Color OpenGL Fast Path Blue option puts a blue overlay over content that is detached from the compositor. The Flash Updated Regions option flashes updated screen regions yellow.

The Core Animation instrument works best for any applications that are using CPU as the rendering mode. If your application is published using the GPU rendering mode, you will see the entire screen area turn blue when the Color OpenGL Fast Path Blue option is selected.

Make sure that you deselect all the options before disconnecting your device. If you do not, all the overlay for the options will still appear on the device.

Using the Core Animation Instrument

Note: *You can perform these steps only on Mac OS X. If you are using Windows, see the section "Using the Hardware Acceleration Profiler" later in this chapter for similar utility.*

① Open Instruments.

② Create a Core Animation instrument.

Note: *See the earlier section "Using Instruments" for more information.*

③ Click here and select an application to profile.

④ Click Record.

• The frames per second are graphed in the Track pane.

• Here is the current frames per second.

⑤ Click a debug option, such as Color Blended Layers.

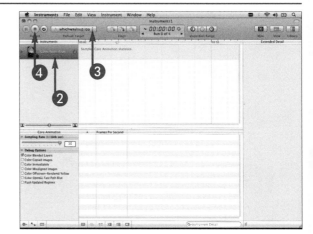

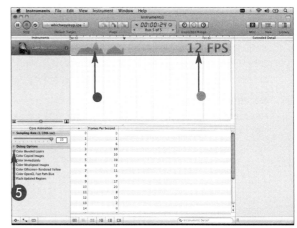

- Green overlays are shown for layers with blending disabled.
- Red overlays are shown for layers with blending enabled.

- Yellow overlays represent areas that have been updated.

Extra

As mentioned earlier, when you are using the Color Blended Layers option, any layers with blending enabled will have a red overlay, and those that do not will have a green overlay. There are a few things that are important about this statement. Layers and images are not the same thing and should not be confused with each other. In the iPhone SDK, you can create layers using the Core Animation framework. These layers can be of any size and are stacked on top of each other just like the layers in the Flash Timeline. Because each layer can take up a specific area of the screen, you can optimize each individual item to be on a separate layer. This makes the redrawing of any assets much more optimized, as you only have to redraw the layer that needs updating.

Because you do not have control or the ability to create Core Animation layers, the Stage is rendered as one layer. If there are any visible graphics that have transparency, a red overlay will be applied to the entire screen. If you were to show a full-screen image without any transparency, a green overlay would be applied.

Using the OpenGL ES Instrument

The Instruments application includes an instrument called *OpenGL ES,* which queries the GPU driver on an iOS device in order to sample OpenGL statistics for your application. Note, again, that Instruments is available only on Mac OS X. Using the OpenGL ES instrument enables you to see how efficiently your application is using OpenGL. In a Flash iPhone application, you do not have direct access to any OpenGL methods or features; however, when your application is published using the GPU rendering mode, your application will use OpenGL to display your assets. For GPU optimization techniques, see Chapter 15, "Optimizing Performance."

Each version of the iPhone and iPod touch ship with different GPU chips, each with a different set of statistics. The Instruments application does not know ahead of time which statistics will be displayed.

The GPU hardware has two components, a tiler and a renderer. Looking at the tiler and renderer utilization values can help in determining performance bottlenecks in your application. Low renderer utilization may mean that the process is waiting for the tiler to complete. Decreasing the complexity of the scene may help resolve this issue. If both the tiler and renderer utilization values are low, it may mean that there is a bottleneck elsewhere in your application.

The Track pane graphs the resource bytes of the OpenGL content. Selecting one of the frames in the Details pane will show the sample's statistics in the Extended Detail pane. The Resource Bytes column in the Details pane displays the number of bytes used for all the textures. You can use this column to make sure that your textures are being removed when they are no longer needed.

Using the OpenGL ES Instrument

Note: *You can perform these steps only on Mac OS X. If you are using Windows, see the section "Using the Hardware Acceleration Profiler" for similar utility.*

Set GPU Rendering Mode

Note: *In order to use the OpenGL ES instrument, you will need to publish your application using the GPU rendering mode.*

1. In Flash, click File → iPhone OS Settings.

   The iPhone Settings dialog box appears.

2. Click here and select GPU.

3. Click Publish and install the application on your device.

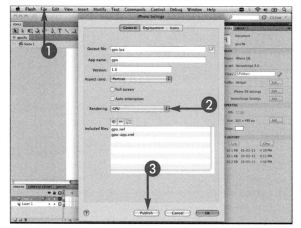

Using the OpenGL ES Instrument

1. Open Instruments.

2. Create an OpenGL ES instrument.

Note: *See the earlier section "Using Instruments" for more information.*

3. Click here and select a hardware-accelerated application to profile.

4. Click Record.

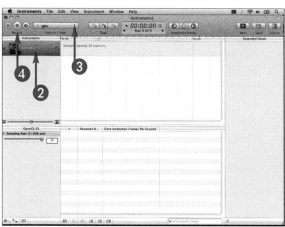

- Resource bytes are graphed by default in the Track pane.
- Resource bytes for each sample are shown in the Details pane.
- Core animation frames per second are shown in the Details pane.

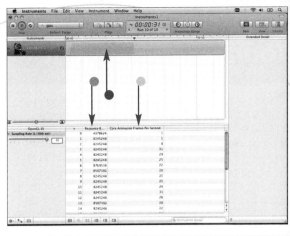

5. Click a sample from the Details pane.
6. Click here.
- The Extended Detail pane shows sample statistics for that sample.

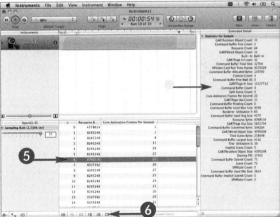

Extra

When optimizing your OpenGL content, the important statistics to keep a close eye on are the total resource bytes and the resource count. Keeping these numbers as low as possible will help you get the best performance out of your application. Also, keep in mind that the Core Animation frames per second value is not the frames per second value of your entire application. This simply measures the animation and not the code execution of your application.

The following are some of the statistics that may be displayed: Device Utilization records any device activity. This accounts for the entire device's utilization, including items such as the Mail application. Context Count returns the number of global OpenGL contexts. It is important to note that there is other processing, other than your application, which may be responsible for the creation of contexts. Resource Count returns the number of textures currently in use. Scene Count returns the number of times a scene was sent to be tiled and rendered. This occurs when the scene is flushed or the command buffer space is filled. The Command Buffer Submit Count statistic is incremented when a command buffer is pushed to the GPU.

Using the Activity Monitor Instrument

The Instruments application includes an instrument called the *Activity Monitor*, which records the load on the device measured against the virtual memory size. Note, again, that Instruments is available only on Mac OS X. The Track pane shows four values by default; however, these can be configured to show just the items whose values you are interested in. VM Size represents the total amount of virtual memory space in use. % Total Load represents the total load on the system. % User Load represents the amount of load generated by the user. % System Load represents the total load generated by the system.

The Details pane shows a data grid of each process currently running on the device. In the Process Name column, you should see the application that you launched to record. The User Name column represents the user that started the process. Any process that is started by the mobile user counts towards the % User Load value. The % CPU column displays the percentage of the CPU that a given process is using. The Threads column represents how many threads the process is currently using. The Real Memory column represents how much memory the process is taking up.

As you interact with your application, the two columns that you want to pay close attention to are % CPU and Real Memory. If the value in the % CPU column gets too high, there is a good chance that your application may be terminated by iOS or may appear to be frozen, which causes the user to force-quit the application. Keeping a close watch on the Real Memory value will give you a better idea if your objects are being properly cleared from memory.

Using the Activity Monitor Instrument

Note: *You can perform these steps only on Mac OS X. If you are using Windows, try Flash Builder's Profiler for similar utility.*

1. Open Instruments.
2. Create an Activity Monitor instrument.

Note: *See the earlier section "Using Instruments" for more information.*

3. Click here and select an application to profile.
4. Click the Info button.

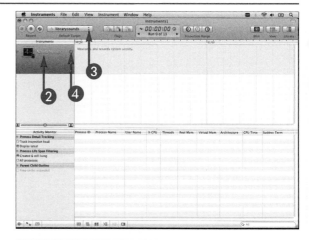

The Info panel appears.

5. Select the system statistics that you want to graph.
6. Click Record.

- The Details pane shows a list of all the processes running on the device.

7 Find the process name that matches the application name.

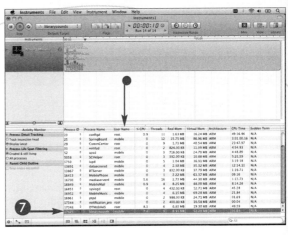

8 Observe the CPU usage as you interact with your application.

9 Monitor the real memory usage as you interact with your application.

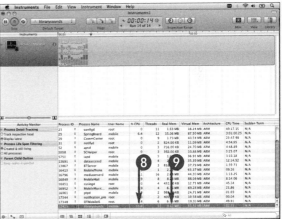

Extra

You can use the I/O Activity instrument, shown earlier in this chapter in the section "Using Instruments," to record any I/O events that occur on files, such as read, write, open, and close. Even if you did not write any code to interact with the file system of the device, your application makes several operations on files in order to function properly. It is also very interesting to see the order in which files get accessed when your application launches. Monitoring how your application is launched may give you some ideas on how to better optimize the code when the application launches, causing it to appear faster. I/O activity will even show when reading and writing items such as a local `SharedObject`.

The Function column specifies the operation being performed on the file. Some functions may be more obvious than others. If there is one that you do not recognize, a simple Google search should bring up its definition. The In Bytes column represents the number of bytes requested to be read or written. The Out Bytes column represents the actual number of bytes read or written to the file.

Using the Hardware Acceleration Profiler

You can use Adobe's Hardware Acceleration Profiler to profile hardware-accelerated applications on the device. Chapter 15 discusses setting your application to use the GPU rendering mode. When your application makes use of this rendering mode, it will create and upload textures of your bitmaps to the GPU. After the textures are uploaded to the GPU, the GPU can be used to perform translations as well as scaling and rotation transformations. The GPU is much faster at performing these tasks than the CPU is.

Depending on the type of translations and transformations that you want to do, you will want to make sure that you are using the `cacheAsBitmap` and `cacheAsBitmapMatrix` methods on the `DisplayObject` class. These will enable you to perform your translations without having the bitmap re-created and re-uploaded to the GPU. For more details on these methods, again, see Chapter 15.

If you are using these methods, the big question is how do you know if your images are being redrawn or not? The Hardware Acceleration Profiler tool, similar to the Core Animation instrument, allows you to see which textures are being redrawn by the GPU by placing a red overlay over the object. When you use this tool, it is important to keep in mind that it will have an effect on the performance of your application. After you are satisfied that you have eliminated any unnecessary redraws, you can turn the tool off and check the performance of your application.

You may also notice that the textures may be a lower quality than before and that all alpha values will snap to either 0 or 1.

To set up your hardware-accelerated application to use the Hardware Acceleration Profiler, you add a key value pair to the `InfoAdditions` node in your application descriptor file.

Using the Hardware Acceleration Profiler

1. In Flash, click File.
2. Click iPhone OS Settings.

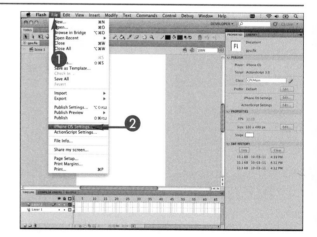

The iPhone Settings dialog box appears.

3. Click here and select GPU.
4. Click OK.

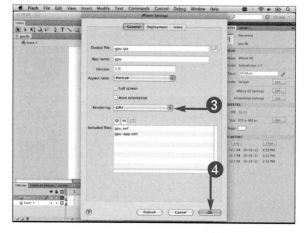

5. Open the application descriptor file.
6. Add an `<iPhone>` node.
7. Add an `<InfoAdditions>` node.

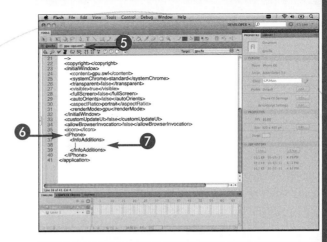

8. Add CDATA, such as `<![CDATA[]]>`.
9. Add the texture tracking key, such as `<key>CTTextureUpload Tracking</key>`.
10. Set the key to `true`, such as `<true/>`.
11. Compile and install the application on a device.

    Textures will have a red overlay when they are being redrawn.

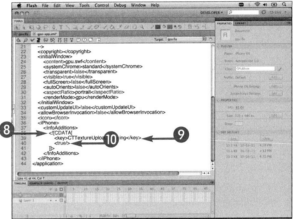

Extra

After you have used the Hardware Acceleration Profiler for some time, you will notice that the texture is uploaded only when it becomes visible on the screen. Even if the display object is added to the display list but is currently off-screen, the texture will not be uploaded until it is shown within the bounds of the screen. This is especially important to know when trying to create scrolling lists. Having each item rendered on-screen first before being laid out will help improve performance.

Here is a quick recap of when to use which methods for GPU optimization. If your 2D display objects are simply going to animate on the x and y, you can use the following syntax in order to have it uploaded to the GPU:

```
myobject.cacheAsBitmap = true;
```

If you want to transform your object by rotating or scaling it, you can use the following syntax to ensure that your texture does not get re-uploaded:

```
myobject.cacheAsBitmap = true;
myobject.cacheAsBitmapMatrix = new Matrix();
```

Create an Application Icon

Every application binary that is uploaded to the iTunes App Store must contain an application icon. This icon will be shown on the home screen of the iOS device and will be used to launch your application on the device. It will also be used in the App Store when viewed on either the iPod touch or iPhone.

The icon should be 57 pixels wide by 57 pixels high and should be completely square. You do not need to add the round corners in your application because iOS will apply a mask to your icon in order to created the rounded corners. The image should be saved as a 24-bit PNG image file format at 72 ppi, and it should be completely flattened and should not contain any transparency.

You should design your icon without the glare embedded in it. The glare is added to the image at the same time that it is masked by iOS. For more details on how to remove the glare from your icon, see the following section in this chapter.

After your icon has been designed, you can add it to your application through the Flash IDE. All icons can be added from the Icons tab in the iPhone Settings dialog box.

When your application is compiled, your icon will be bundled with your application and renamed to Icon.png. When your application is installed, iOS looks for an icon with this name inside your bundle.

Create an Application Icon

1. Open an image-editing application, such as Photoshop.
2. Create a 57 x 57 square icon.
3. Click File → Save As and save the icon as a nontransparent PNG file.

4. In Flash, click File.
5. Click iPhone OS Settings.

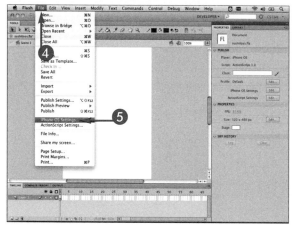

The iPhone Settings dialog box appears.

6 Click the Icons tab.

7 Click the icon 57 x 57 item.

8 Click the folder button and select your image.

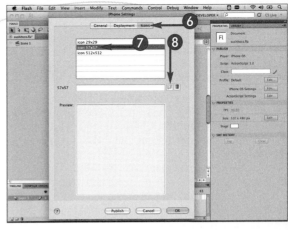

● Your icon will appear in the Preview area.

9 Click OK.

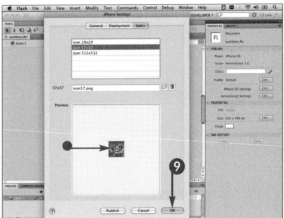

Chapter 18: Deploying Your Application

Extra

It is a good idea to design your icon at a bigger resolution than the 57 x 57 resolution. This will allow you to create higher quality graphics later. Scaling down your icon to fit the required dimensions will be easier than scaling it up later. If your larger icon contains a lot of detail, you may find that it is lost when it gets scaled down. In this case, it may be okay to create a smaller version of your icon.

There are a few things to keep in mind when designing your icon. Using bright and bold colors will make your icon stand out in the App Store as well as on a device. Also, keep in mind that the background color of the home screen on the device is black, and the background of the App Store is white. Creating an icon that is primarily black or white may cause it to be hard to find. Also, keep in mind if you plan on releasing your application in App Stores worldwide, text may not be a good solution. Designing something that looks like Apple designed it goes a long way in making your application stand out.

307

Remove the Glare from Your Application Icon

When you create your icons, it is important that you do not add the rounded corners and the glare to them. These are added to your icon by iOS, so if you added them, they would conflict with the glare added by the iPhone. If, for some reason, you do want to remove the glare that the iPhone puts on your icon, you can. This enables you to add your own effect; however, it is strongly recommended that if you want a shine or gloss on your icon, you should keep the same effect that Apple has provided.

The `UIPrerenderedIcon` key in the Info.plist file is a Boolean value that can turn the shine on or off. This does not have any effect on the rounded corners of your icon, and your icon will always be masked. The value for this key is also applied to the large application icon that is used in the iTunes App Store. For more details on this, see the section "Create App Store Graphics" later in this chapter.

The Info.plist file is a configuration file for your iPhone applications that contains many properties describing your application. Because this file is generated by the Flash iPhone Packager when your application is compiled, you cannot add the key directly to the file. You can add any additional information you want to have included in the Info.plist file in the application descriptor file for your project. The `<InfoAdditions>` node, which is not present by default, is used to define any additional keys to be added. Placing the key value pairs in a CDATA tag allows the Flash compiler to ignore these nodes and use them only for inserting into the Info.plist file.

Remove the Glare from Your Application Icon

① Open the application descriptor file.

② Add the `<iPhone>` node.

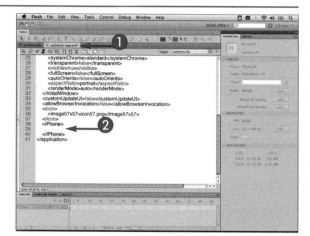

③ Add the `<InfoAdditions>` node.

④ Add a CDATA node, such as
`<![CDATA[ ]]>`.

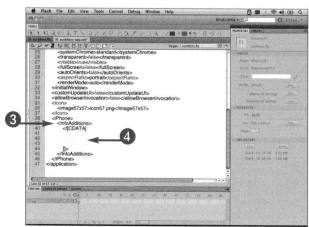

5 Add the `UIPrerenderedIcon` key, such as `<key>UIPrerenderedIcon</key>`.

6 Set the key to `true`, such as `<true/>`.

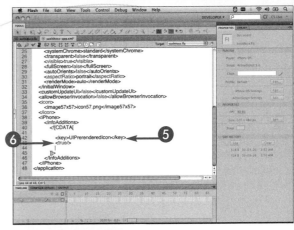

7 Compile and install the application on your device.

● This is the icon with the glare.

● This is the icon without the glare.

Extra

When you are adding additional elements to the Info.plist file, it is extremely important that you make sure that the syntax is correct. Your application will still compile and could even install correctly even if there is an error. However, if there is an error, you will run into problems when trying to submit your application to the App Store through the iTunes Connect portal. Always make sure to test every feature of any of the additional entries. You can also open the Info.plist file with the Property List Editor application on Mac OS X. If there is a problem with the structure or syntax of the XML, you will be able to see it in the application, or it will alert you of the error.

To open the Info.plist file, you will need to uncompress your .ipa file. The IPA file format is simply a ZIP file format. Renaming the file extension to .zip will enable you to uncompress the file just like any other ZIP file. If you are using a Mac OS X machine, you will find the .app file in the Payload folder that is extracted from your ZIP file. Right-click the .app file and click Show Package Contents in order to see the application bundle. Find your Info.plist file and double-click it to open it.

Create a Default Splash Screen

When the user launches your application, the entire binary of your application is loaded into memory. Applications created with the Flash iPhone Packager can grow in file size quickly. The bigger the application is, the longer it will take to be loaded into memory. The default behavior is to show a black screen as the application loads. This is not ideal as it gives the user the impression that the application is not working, especially if the application load time is long.

You can, however, show a default splash screen as your application loads in the background. To accomplish this, simply bundle a Default.png file in the root directory of your application. If iOS detects the presence of this file, when launched, the default splash screen will be displayed to the user. Also, the image will be animated as the application launches.

The image dimensions of the Default.png file should fill the entire screen. If the status bar is visible, it will be placed over the top of the image. Also, if your application supports only one screen orientation, you should make sure that the image is created in that mode. This will give the user a hint as to which orientation the device will launch in. After the application has finished loading, the Default.png image is removed, and your application will be shown.

If you plan on doing any additional loading when your application begins, you can include the image in your application as well and make it the first thing that is shown when the application launches. This will allow you to show any additional progress indicators, alerting the user that more information is being loaded.

Create a Default Splash Screen

1. Open an image-editing application, such as Photoshop.
2. Create a full-screen splash screen image.
3. Click File → Save As and save the image as Default.png.

4. In Flash, click File.
5. Click iPhone OS Settings.

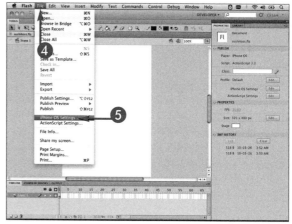

The iPhone Settings dialog box appears.

⑥ Click the General tab.

⑦ Click the + button and add your file in the Open dialog box.

Note: Select the Default.png image that you created in step **2**.

● Your splash screen will now be bundled with your application.

⑧ Click Publish.

When you compile and install your application on a device, you will see the splash screen.

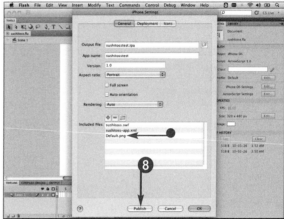

Extra

One method that some applications use for the Default.png image is using a screenshot of the very first state of your application. There are several ways you could get a screenshot of your application. While testing it in the Flash IDE, you could take a screenshot and crop the image in an image-editing application such as Photoshop. You could also write some temporary code in your application to take a bitmap snapshot of the first screen and save it out to a PNG image on your local hard drive. One of the easiest ways is to take the screenshot when the application is running on the device. The iPhone and iPod touch have the ability to capture and save an image of the entire screen. The image is then added to the Camera Roll in the Photos application. The easiest way to get the image off the device is to email it to yourself. Selecting the image in the Camera Roll will present you with a button to send the photo. When the alert sheet appears, select email. One application that uses this technique well is the Facebook application.

Create App Store Graphics

After you have developed your application, you will need to do a few things in order to get it ready to be submitted to the iTunes App Store. The iTunes page for your application will contain a large icon and screenshots of your application.

The large application icon should be 512 pixels wide and 512 pixels high. Make sure that the icon is completely square and does not contain any rounded corners; Apple will apply a mask to your icon in order to create the rounded corners. The image should be a 72 ppi, RGB, high-quality JPEG or TIFF image. It should be completely flattened and should not contain any transparency. iTunes Connect does not support PNG images for your large application icon. Also, ZIP-compressed TIFF images are not supported and will not appear correctly in the App Store. It is a good idea to create this icon at the proper resolution. Scaling up smaller graphics to the correct size can cause the image to appear pixilated and blurry.

If you have submitted your application with the glare removed from the icon, it will be reflected with the large application icon as well.

Submitting screenshots for your application will allow the user to see your application before buying it. For best results, do not include the status bar of your device in the screenshots. You can supply both portrait and landscape mode screenshots. Screenshots in portrait mode will be displayed in both the App Store in iTunes and the App Store on the device, whereas landscape mode screenshots will be shown only in iTunes. You can add up to four additional screenshots of your application.

Create App Store Graphics

Create the Large Application Icon

1. Open an image-editing application, such as Photoshop.
2. Create a 512 x 512 application icon.
3. Click File → Save As and choose JPEG as the format.

The JPEG Options dialog box appears.

4. Click here and set the quality to Maximum.
5. Click OK.

Your large icon is now ready to be submitted with your application.

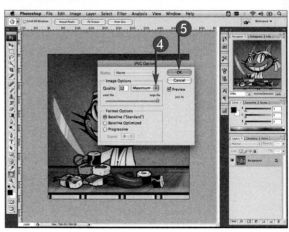

312

Create Screenshots

Note: You can take screenshots on your device by pressing the lock and home key at the same time.

1. Open an image-editing application, such as Photoshop.

2. Open the primary screenshot of your application.

3. Click File → Save As and choose JPEG as the format.

The JPEG Options dialog box appears.

4. Click here and set the quality to Maximum.

5. Click OK.

You can create an additional four screenshots of your application by following steps **2** to **5** for each one.

Your screenshots are ready to be submitted with your application.

Extra

After you have submitted your application and are waiting for it to be approved, it is a good idea to take the time to get some additional artwork together. There is always the potential that someone from the Apple Worldwide Developer Relations team will contact you about featuring your application in the App Store. And if they come calling, you will want to have all the material ready for them. Being a featured application in the store greatly enhances your sales and downloads. The additional work can be different than your application icon and screenshots and should be designed as promotional material.

Currently, there are two different featured graphic sizes on the home page of the App Store. The large banner at the top is approximately 800 pixels wide by 300 pixels high, and the smaller banner to the right is approximately 200 pixels wide and 100 pixels high. When you are creating these graphics, keep them as layered Photoshop files. This will make it easy for you or the Apple team to change them in order to fit any additional sizes.

Create a Distribution Certificate

Throughout the development of your application and throughout this book, you have been using a development certificate to sign your application. When your application is ready for other users to try out, you will need to create a distribution certificate. This enables you to digitally sign your application so that you can distribute ad hoc builds, as well as upload your application to the App Store. It is important to note that only team agents are allowed to create distribution certificates.

The process for creating a distribution certificate is very similar to the process for creating the development certificate. The first step is to create a certificate signing request. This process is the same as it was for the development certificate. To refresh your memory, see the section in Chapter 1, "Getting Started with iPhone Development," about generating a CSR for your particular operating system.

After you have created your request, you can upload it to Apple through the iPhone Developer Program Portal. After the certificate signing request has been uploaded to Apple, you will be asked to approve the certificate. After you click the Approve button, you can refresh your browser, and your certificate will be available to download. Keep in mind that all certificates have an expiry date and will need to be updated.

Create a Distribution Certificate

1. In a Web browser, go to the iPhone Developer Program Portal Web site, http://developer.apple.com/iphone/manage/overview/index.action.

2. Log into the portal.

   The Welcome page appears.

3. Click Certificates.

   The Certificates tabs appear.

4. Click the Distribution tab.

5. Click Request Certificate.

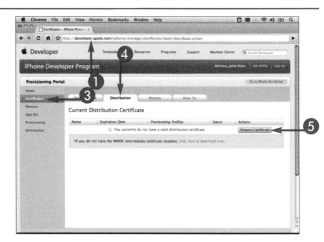

   The Create iPhone Distribution Certificate page appears.

6. Click Choose File.

   A dialog box appears.

7. Navigate to and click your CSR file.

8. Click OK.

   You are returned to the Create iPhone Distribution Certificate page.

9. Click Submit.

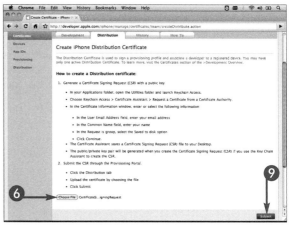

Your certificate signing request is uploaded to Apple.

10 Click Approve.

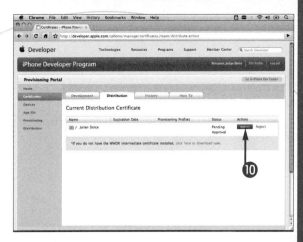

11 To download your certificate, click Download.

Note: Export your certificate to a P12 certificate after it is downloaded to your file system.

The distribution certificate can be used to compile distribution applications.

Extra

After you have downloaded the distribution certificate from Apple, you will need to export a .p12 file from it. This is the same type of file you exported from your development certificate and is used by Flash to digitally sign your applications.

The process for exporting the P12 certificate is different on Mac OS X and in Windows. If you are developing on an OS X machine, it is fairly straightforward and can be done using the Keychain Access application. Double-clicking your certificate file will install it in the Keychain Access application. After it is installed, you can right-click the certificate and select Export from the context menu. The on-screen instructions will walk you through the process of saving the .p12 file and creating a password.

Creating the P12 certificate from your distribution certificate is much more involved when developing on a Windows machine. There are several steps that require some command-line work. For more detailed instructions for creating a P12 certificate, see Chapter 1.

Create an Ad Hoc Provisioning File

During the development of your iPhone application, you have used a development provisioning profile file in order to install your application on your device. The provisioning profile file connects your device with your iPhone developer account, which allows you to install the application on it for testing.

When you are ready to send your application to friends or people in your company, in order to test, you can create an ad hoc provisioning profile. This allows other users who are not developers to install the application on their device.

The one thing you will need to do is get all the device UDIDs from your users and add them to your iPhone developer account. Keep in mind that you can add only 100 devices a year to your account. For more details on how to add devices to your account, see Chapter 1. It is a good idea to wait until you have received all the device UDIDs from your users before creating an ad hoc provisioning profile, as you cannot add devices to a profile after it has been created.

You can use the iPhone Developer Program Portal to create your ad hoc provisioning profile.

Create an Ad Hoc Provisioning File

1. In a Web browser, go to the iPhone Developer Program Portal Web site, http://developer.apple.com/iphone/manage/overview/index.action.

2. Log into the portal.

   The Welcome page appears.

3. Click Provisioning.

   The Provisioning tabs appear.

4. Click the Distribution tab.

5. Click New Profile.

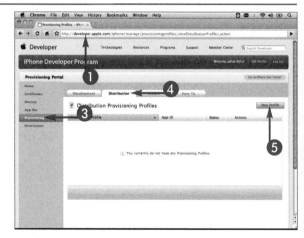

   The Create iPhone Distribution Provisioning Profile page appears.

6. Click Ad Hoc.

7. Give your profile a name, such as ad hoc profile.

8. Click here and select an App ID.

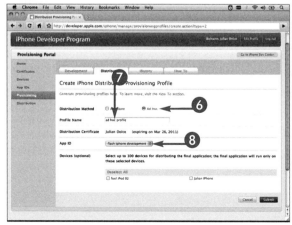

9 Click all the necessary devices in the list.

10 Click Submit.

You are returned to the Distribution Provisioning Profiles page.

11 To download the profile to your file system, click Download.

The ad hoc profile is ready to be used to publish an ad hoc distribution application.

Chapter 18: Deploying Your Application

Extra

For some people, finding the UDID of their device may not be any easy task. For those who are not very technologically savvy, the process may be intimidating. There are a couple of great applications in the iTunes App Store that make it very easy to send an email with the ID prefilled in the body of the email. Ad Hoc Helper by Erica Sadun does an excellent job of this. It has been my experience when dealing with a lot of testers that you should send them instructions on downloading this application to email you their UDIDs. Pretty much every user will know how to search the store, either on their device or in iTunes, to find and install an application. It may be overwhelming to receive a lot of device UDIDs all at the same time, and adding them to your account can be painfully slow. You can use the iPhone Configuration Utility application to export all the UDIDs in an XML file. Or you can create the XML file yourself as you receive a new device. For more details on how to create a .deviceids file to do a bulk upload, see the Apply It area of the "Add Devices to Your Account" section in Chapter 1.

Publish for Ad Hoc Distribution

After you have received all the Device UDIDs from your testers and created your distribution certificate and ad hoc provisioning profile, you can compile your application for ad hoc distribution. The process for publishing your application for ad hoc distribution is very similar to publishing it for development. You will, however, have to make some changes to the publishing settings of your project in order to properly publish for ad hoc distribution.

The iPhone Settings dialog box's Deployment tab contains all the settings that need to be changed. In the Certificate section, you select the distribution certificate you created earlier. After you have added it to any file in Flash, you can select it again from the drop-down list. You will need to enter the password that was set when creating your P12 certificate. After you have added the ad hoc provisioning profile that you created and downloaded from the iPhone Program Portal to a file in Flash, you will be able to select it from the Provisioning Profile drop-down list.

When you created your ad hoc provisioning profile, you selected an App ID to be associated with it. You will need to enter that App ID.

After all your ad hoc distribution credentials are set and you choose ad hoc deployment, your application will be ready to be published and distributed to your users. When sending the application to your testers, do not forget to include the ad hoc provisioning file as well. Each user will have to install both on his or her device in order to test the application.

Publish for Ad Hoc Distribution

1. In Flash, click File.
2. Click iPhone OS Settings.

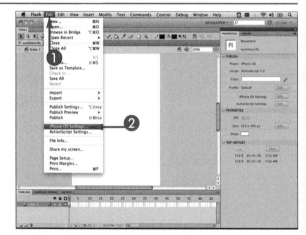

   The iPhone Settings dialog box appears.

3. Click the Deployment tab.
4. Click here and select your distribution certificate.
5. Type your certificate password.
6. Click Remember Password for This Session.

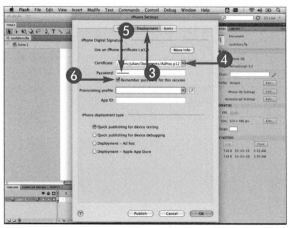

318

7 Click here and select your ad hoc provisioning profile file.

8 Type the App ID associated with your profile.

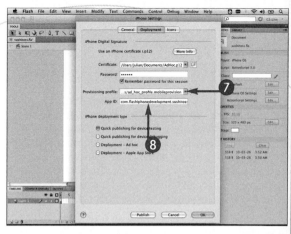

9 Click Deployment – Ad Hoc.

10 Click Publish.

The application is published for ad hoc distribution.

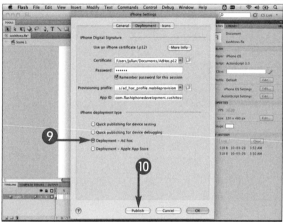

Extra

If you have been using the command line in order to compile your applications, you will have to change the development credentials in your script to match those of your distribution credentials. You will also need to change the -target switch in order to have your application be compiled for ad hoc distribution. Here is an example of the new command:

```
java -jar pfi -package -target ipa-ad-hoc -provisioning-profile "adhocprovisioning.
 mobileprovision" -storetype pkcs12 -keystore "adhoccertificate.p12" -storepass password
 "settings.ipa" "settings-app.xml" "settings.swf" "Default.png" "appicon.png"
```

Create an App Store Provisioning File

Before you can publish your application to be submitted to the App Store, you will need to create a provisioning profile for the App Store. You do not have to create a new certificate because you can use the same distribution certificate that you created for ad hoc distribution. If you have not already created a distribution certificate, see the section "Create a Distribution Certificate" earlier in this chapter.

If you have also been using a wildcard-style App ID for your development, this may be a good time to create a new one specifically for your application. Wildcard App IDs should still work; however, because Flash CS5 iPhone applications do not have the functionality to take advantage of this, it is a good idea to create an app-specific one for distribution.

It is important to note that only a team agent can create all types of distribution provisioning files, including both App Store and ad hoc profiles. If you are not the team agent for your developer account, ask him or her to create one for you.

If you are the team agent, you can use the iPhone Developer Program Portal to create your App Store provisioning profile. You will use the Provisioning settings. By default, the Development tab is selected, so you will need to select the Distribution tab to create a distribution profile. When you are filling out the information about your profile, be sure to select App Store as the distribution method. It is also a good idea to put "app store," or something similar, in the name of your profile in order to distinguish it from other files later.

Create an App Store Provisioning File

1. In a Web browser, go to the iPhone Developer Program Portal Web site, http://developer.apple.com/iphone/manage/overview/index.action.

2. Log into the portal.

   The Welcome page appears.

3. Click Provisioning.

   The Provisioning tabs appear.

4. Click the Distribution tab.

5. Click New Profile.

   The Create iPhone Distribution Provisioning Profile page appears.

6. Click App Store.

7. Type a name for your profile, such as sushi toss test app store.

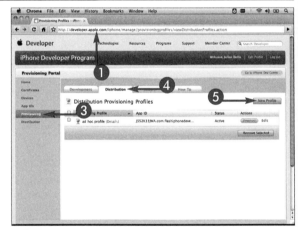

320

⑧ Click here and select the App ID for the application that you are going to submit to the App Store.

⑨ Click Submit.

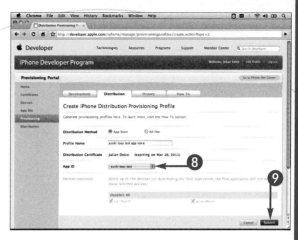

You are returned to the Distribution Provisioning Profiles page.

⑩ To download your App Store profile to your file system, click Download.

The provisioning profile is ready to be used to publish an application for the App Store.

Extra

If, at any point in time, you need to modify your provisioning profile, you can do so using the iPhone Program Portal. For each profile in the list, there is an Edit button at the far right of the row. Click this button to bring up a second context menu with Modify and Duplicate options. Modifying the existing profile enables you to make any necessary changes — for example, removing devices or changing the name, App ID, or even the distribution type. Duplicating profiles can be a handy way of getting a new profile started and completed quickly. There is a good chance that your ad hoc distribution provisioning profile will be very similar to the App Store distribution provisioning profile, so duplicating the ad hoc profile enables you to have to change only the distribution type and the name of the profile.

Editing any profile requires it to be approved, redownloaded, and reinstalled on every device. If it is named the same, this can be slightly confusing when you are trying to determine if you have the most up-to-date version of the file installed on your device. Incrementing a number in the name can help eliminate these problems.

Publish Your Application for App Store Distribution

This is the moment you have been waiting for. You have finished development of your application and tracked down and squashed all the bugs that your testers have found. You are now ready to publish your application for submission to the iTunes App Store. To successfully publish your application, you will need to have created a distribution certificate and an App Store provisioning profile. If you have not created either one of these, see the sections "Create a Distribution Certificate" and "Create an App Store Provisioning File" earlier in this chapter.

You will use the iPhone Settings dialog box to make the necessary changes in order to publish for the App Store. The General tab contains some basic information about your application. Make sure that the app name is the same name that you want to have shown in the App Store. Also, make sure that the version is correct. If this is the first time you are going to submit your application, you should set the version to 1.0 or something similar.

On the Deployment tab, you will set the distribution credentials in order to publish for the App Store. In the Certificate selection box, make sure that your distribution is selected. You will also need to enter the password that you created when exporting the P12 certificate.

When you select your App Store provisioning profile, the App ID may automatically change for you, but you should always double-check that it is the same App ID that is associated with your profile.

Publish Your Application for App Store Distribution

① In Flash, click File.

② Click iPhone OS Settings.

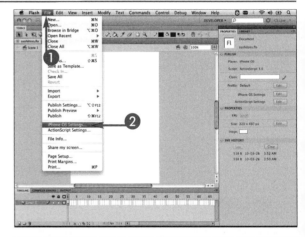

The iPhone Settings dialog box appears.

③ Click the Deployment tab.

④ Click here and select your distribution certificate.

⑤ Type your certificate password.

⑥ Click Remember Password for This Session.

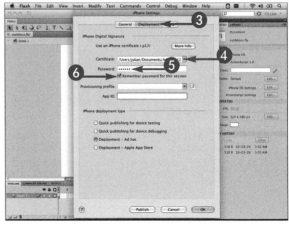

322

7 Click here and select your App Store provisioning file.

8 Type the App ID for your application.

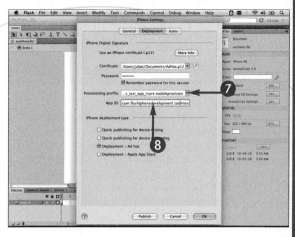

9 Click Deployment – Apple App Store.

10 Click Publish.

Your application will begin publishing.

11 When your application has finished publishing, rename the .ipa file to a .zip file.

Note: *This is the file that you will upload to Apple.*

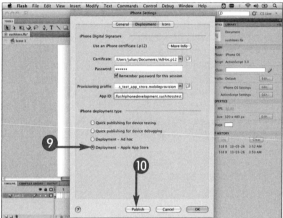

Extra

If you have been using the command line to compile your applications, you will have to change the development credentials in your script to match those of your distribution credentials, as mentioned earlier. You will also need to change the -target switch in order to have your application be compiled for App Store distribution. The following is an example of publishing for the App Store:

```
java -jar pfi -package -target ipa-app-store -provisioning-profile "appstoreprovisioning.
  mobileprovision" -storetype pkcs12 -keystore "distributioncertificate.p12" -storepass password
  "settings.ipa" "settings-app.xml" "settings.swf" "Default.png" "appicon.png"
```

Submit Your Application to the App Store

After you have successfully published your application for the App Store, you are ready to upload it through the iTunes Connect portal. Along with your application, you will be uploading all the necessary artwork, such as screenshots and application icons. If you plan on charging for your application, you will need to have filled out the required contracts and banking information, which can be found on the iTunes Connect site.

After you have logged into the iTunes Connect portal, you can start the process of adding your new application. You may be asked to select the primary language and the company name for your application. If so, the company name cannot be changed later, so be sure to enter the correct name. The next screen is Export Compliance. There is a good chance your application is not using encryptions; if that is the case, you can skip this section.

The Overview tab of the Add New Application page is where you enter all of the metadata associated with your application. After you have submitted this information, you will not be able to edit it again until you submit an update to your application — or if Apple or the Developer, as Apple will refer to you, rejects the application. Filling these fields out correctly is extremely important to the success of your application. The Application Name and Keywords fields will be used when a user performs a search in the App Store. Also, the Application Description field's information will be the key piece of text users read when deciding to purchase or download your application. Take your time filing this form out. Do not rush to submit your application.

Also, make sure that the application name and version match those in the iPhone Settings dialog box in the Flash IDE.

Submit Your Application to the App Store

1. In a Web browser, go to the iTunes Connect site, http://itunesconnect.apple.com.

2. Log into your iTunes Connect account.

   The home page appears.

3. Click Manage Your Applications.

   The Manage Your Applications page appears.

4. Click Add New Application.

   The Export Compliance page appears.

5. Fill out the form.

6. Click Continue.

   The Add New Application page appears.

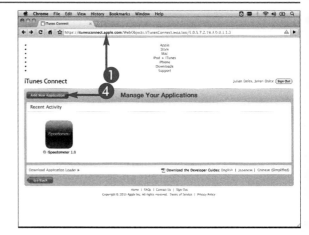

7. Type your application name.

8. Type the application description to appear on the product page.

9. Choose whether you want to have your application limited to certain devices.

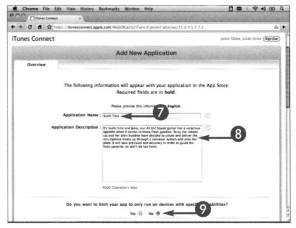

324

⑩ Scroll down on the page.

⑪ Click here and select a primary category, such as Games.

⑫ Select subcategories of your primary category.

⑬ Click here and select a secondary category, such as Entertainment.

⑭ Type your copyright information.

⑮ Type your version number.

⑯ Type a SKU number.

Note: *The SKU number can contain both letters and digits.*

⑰ Scroll down on the page.

⑱ Type keywords for your application, separating the words with commas.

⑲ Type the application support URL.

⑳ Type the application support email address.

㉑ Scroll down on the page and click Continue.

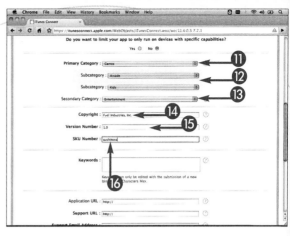

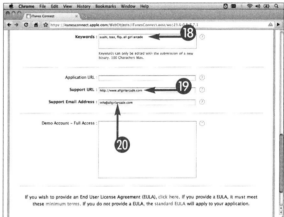

Apply It

One of the questions in the Overview form asks if your application should limit which types of devices your application can be installed on. For example, if your application is primarily a GPS application, you will not want iPod touch users to download it. If this is the case, you will want to add the `UIRequiredDeviceCapabilities` key to your application descriptor. There is a complete listing of all the valid key types for this key on the Apple Developer Web site, http://developer.apple.com/iphone/. Here is an example of what you would add to your application descriptor:

```
<iPhone>
            <InfoAdditions>
              <![CDATA[
                  <key>UIRequiredDeviceCapabilities</key>
                  <dict>
                      <key>gps</key>
                      <true/>
                  </dict>
              ]]>
            </InfoAdditions>
</iPhone>
```

continued

Submit Your Application to the App Store (continued)

Each application in the App Store gets an app rating, similar to an ESRB rating for video games. This rating is generated for you based on a series of questions about the content of your application. This is done for parental controls in the iPhone App Store.

The Upload page is where you will upload your application and the assets of your application. When you are uploading your application binary to Apple, the iTunes Connect portal is expecting a ZIP file. In order to upload your application, simply rename the .ipa file created by the Flash IDE to .zip. When your application is uploaded, it is checked by Apple to ensure that it is a valid application. If the upload succeeds, you will see a check mark graphic appear in the Application box. Any errors will be displayed at the top of the HTML page. If your application was uploaded unsuccessfully, try recompiling your application.

You can also choose to upload your binary at a later time if you prefer. If you are having issues uploading, you can select this option and continue the submission process. There is an application called *Application Loader* that can be downloaded from the iTunes Connect portal that will allow you to upload your binary without going through the submission process again. The Application Loader is installed to the /Developer/Applications/Utilities folder and is available only for Mac OS X.

Next, submit the 512 x 512 large application icon graphic and the screenshots for your application. For more details on the specification and format of this graphic, see the section "Create App Store Graphics" earlier in this chapter.

The remaining pages of the submission process have you set the pricing and localization of your application.

Submit Your Application to the App Store (continued)

The Ratings tab of the Add New Application page appears.

㉒ Answer all the ratings questions.

• Your app rating is generated and appears here.

㉓ Click Continue.

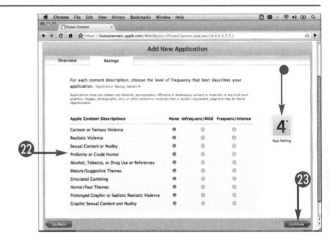

The Upload tab of the Add New Application page appears.

㉔ Click Choose File.

㉕ Choose the .zip version of your application.

㉖ Click Upload File.

Your application file is uploaded.

326

㉗ Click Choose File under Large 512 x 512 Icon.

㉘ Choose your large application icon JPEG.

㉙ Click Upload File.

Your application icon file is uploaded.

㉚ Click Choose File in the screenshots area.

㉛ Choose your primary screenshot file.

㉜ Click Upload File.

Your screenshot is uploaded.

Note: *You can repeat steps **30** to **32** for each additional screenshot, if you have more.*

- This check mark indicates that the application binary uploaded successfully.
- This check mark indicates that the large application icon uploaded successfully.
- This check mark indicates that the screenshots uploaded successfully.

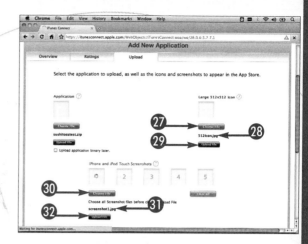

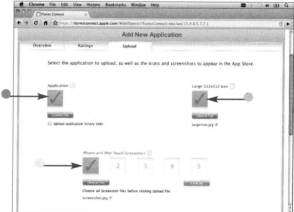

Extra

There is a good chance that the first time you attempt to upload your application, you will receive an error. There could be several reasons for this, and Apple does not provide very specific error messages. The first thing to try is to simply recompile your application. Also, double-check to make sure that your distribution credentials are correct in the iPhone Settings dialog box and that the application name and version match what you entered in the submission process. If you are still having issues, delete your distribution certificates and App Store provisioning profiles and re-create them. In most cases, this will usually solve the problem.

In addition, check the contents of your application bundle. Make sure that the Default.png splash screen is present along with the Icon.png and Icon-small.png image files. The omission of any one of these files could cause the application to be rejected.

If you receive an error stating that the Info.plist file could not be parsed or is missing, there is a good chance that you have included additional parameters that are not correct. Make sure that all additional Info.plist keys in the `<InfoAdditions>` node in the application descriptor are correct.

Getting Your App Approved

After your application has been submitted to Apple, it goes through a review process by a member of the Apple Application Review team. One of the biggest reasons an application is rejected is if, on first load, it crashes for the reviewer. A crashing application is a quick way to have your application rejected.

When filling out your keywords, do not use words that are unrelated to your application. Also, trademarked words and buzzwords are frowned upon by Apple and should be avoided.

If you have created or used graphics that are similar to the iPhone SDK graphics, make sure that they are used as intended. For example, the small "i" button, which can be found in the Weather application, should be used in similar utility type apps and show application preferences.

Using this graphic to show any other information, such as credits, may cause your application to be rejected.

Another big reason for application rejection is the use of private frameworks. Flash iPhone applications do not have the ability to use these private frameworks; however, accessing any of the files outside of the application sandbox could cause your application to be rejected.

When developing iPhone applications with Flash, it is hard to get into too much trouble, as a lot of the main rules are not available for you to abuse. If you develop honest content and follow development best practices, you should have no problems getting your application accepted. If your application does get rejected, it will most likely be something small that can be fixed and resubmitted.

View Approval Status

After you have submitted your application, you can view its approval status.

- Here is your application's approval status.

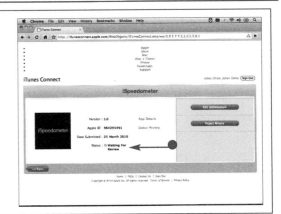

Use Icons As Intended

The info icon should always be used to flip a utility application to show its preferences.

- This is the info icon.

Avoid Improper Keywords

Do not use buzzwords or brands as keywords. Also, do not use words that do not apply to your application.

- These words would be unacceptable because they are buzzwords.
- This is unacceptable because it is a brand name.
- This is a poor choice because it does not relate to my application.

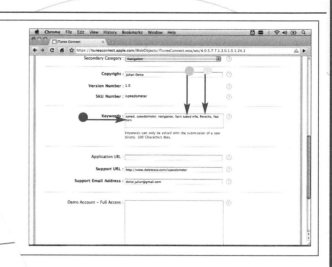

Avoid Reading Files Outside the Sandbox

Do not access file locations, such as the root of the device, that are outside of the application file sandbox.

- This file is outside the application file sandbox, so it should not be accessed.

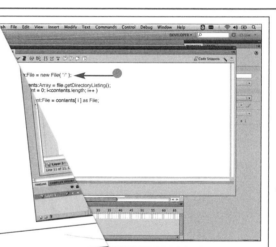

Test for Being Offline and Interruptions

If your application uses any sort of network connection, make sure to test how your application responds if there is not a network connection present or if the application is in Airplane mode. Make sure that you show the appropriate error messages and that the application does not hang if the device is offline.

Also, make sure to test how your application performs when the user receives a phone call, text message, or other interruption. Does it exit gracefully? What happens when the application relaunches when the interruption is finished? You should always save the state of the application often in case this occurs, so you can reload the last known state when the application relaunches.

Make a Revi...

The application...
of days. Some...
approvals, and o...ss can take any number
find that your app... experienced very quick
approved, get the ... wait months. If you
seen a number of T... long time to be
by somebody at App... possible. I have
reviewer there, it will ...ages replied to
application approved q... touch with a
...tting your

Chapter 18: Deploying Your Application

Track Your Application Sales

The day your first application is approved and released in the store is a very exciting day. By day two, you will be anxiously awaiting to see the download numbers of your application, especially if your application is not free. There is a section in the iTunes Connect portal that enables you to generate reports to see the sales and download figures of your application.

There are three different types of transaction summary reports that can be generated: Monthly, Weekly, and Daily. After you select a time period, you can specifically set for which days you would like to see data. You have the option to simply preview the report on the Web site or have it downloaded to your computer. If you choose to download the report, an archived tab-delimited text file is created. The report is hard to read if opened in a regular text editor; however, opening it in a spreadsheet application such as Microsoft Excel or Apple Numbers will separate the data into rows and columns.

The report will show all the application downloads in every App Store worldwide. It will also identify if the download was a purchase or if it was an update. It will further show if the application was downloaded with a promo code. If you have multiple applications submitted to the App Store, the reports will have data for each one, as they are not a separate report for each application.

It is important to note that reports are purged on a regular basis, so you should always download the reports to your computer regularly.

Track Your Application Sales

1. In a Web browser, go to the iTunes Connect site, http://itunesconnect.apple.com.

2. Log into your iTunes Connect account.

   The home page appears.

3. Click Sales and Trends.

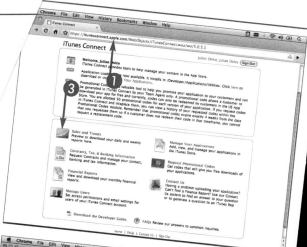

   The Report Options page appears.

4. Click here and select a report type, such as Summary.

5. Click here and select a report period such as Weekly.

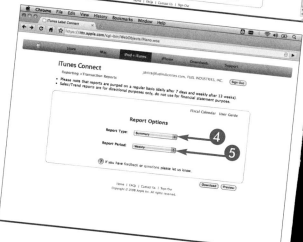

Another drop-down list appears, with options for the report period.

6 Click here and select a timeframe, such as a specific week.

7 Click Preview.

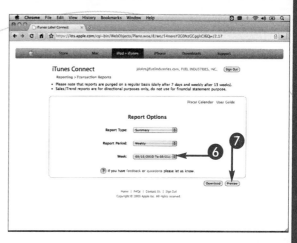

- The Artist/Show column displays your company name.
- The Title/Episode/Season column shows the application title.
- Product Type Identifier 1 means a download or purchase.
- Product Type Identifier 7 means an application update.

8 Click Download at the top or bottom of the page.

The report is downloaded to your computer.

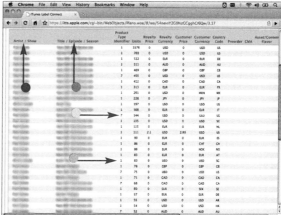

Extra

There are a few open source projects that make downloading the sales reports easier. The open source appdailysales project of Google Code, available at http://code.google.com/p/appdailysales/, is a Python script that downloads the daily sales reports from your iTunes Connect account. This can be extremely useful if set up on a cron job, or scheduled task, that can automatically run every day in order to get the reports. After the reports are downloaded, you can create a script to save the data to a database. This will enable you to generate your own reports, which can be more specific than the ones given by iTunes. For example, you would be able to see that Tuesday is usually the slowest day of the week for application sales and downloads.

There are a few applications in the App Store, such as iSales by PunkStar Studios and Sales Report by Marigo Holdings, that will also give you better insight into your sales data. If you do a search for iTunes Connect in the App Store, you will find a few of these applications. They are relatively expensive applications, but they provide you with invaluable data, which you can use to generate more sales.

INDEX

A

abc (ActionScript byte code), 2
accelerometer, 6, 100-107, 110-111
ActionScript
  coding, 2, 30, 32
  errors, 63
  iPhone Packager with, 9
  `MovieClip` with, 36-37
  objects, 42
  settings, 34
  text layout, 170
  vectors, 48
  with video, 147
ActionScript byte code (abc), 2
ActionScript Message Format (AMF), 174
Activity Monitor instruments, 302-303
ad hoc distribution, 318-319
ad hoc provisioning profiles, 316-318
`addChild()`, 42, 43
`addEventListener()`, 46
address books, 218
AdMob, 248-249
Adobe
  ActionScript Message Format (AMF), 174
  AIR 1.0/2.0, 7, 8-9, 171, 205
  Flash. *See* Flash
  Help, 50-51
  Media Encoder CS5, 148, 149
  Soundbooth, 131
ADPCM files, 132
advertising, 242, 244-245
AIR 1.0/2.0, 7, 8-9, 171, 205
Airplane mode (WiFi), 226, 227
alpha in images, 40, 112, 250
Amazon Kindle, 171
AMF (ActionScript Message Format), 174
analysis of applications, 246-249
animation, 36-39, 52, 100-101, 116
APIs, 6-7, 8-9, 205
.app files, 68, 70, 71, 309
App IDs, 24-25, 60
App Store
  Adobe control over, 8
  applications, rejection, 74, 151
  distribution policies, 10
  graphics, 312-313
  inside applications, 9
  iTunes, 5
  marketing in, 307
  provisioning profiles for, 320-321
  publishing for, 322-323
  requirements for, 306, 309, 314
  resources section, 17
  submissions, 324-327
  version numbers, 72
  Web site banners, 313
appending files, 204-205
AppEngines (font displays), 162
Apple. *See also specific applications; specific devices*
  App IDs generated by, 24-25
  application rejection, 126, 128
  application review, 328
  application user interface standardization, 74-75
  developer registration, 10
  iPhone. *See* iPhone
  iTunes. *See* iTunes
  sending to, 16
  support from, 17
  Worldwide Developer Relations team, 313
Application Loader, 326
applications. *See also specific applications*
  advertising with, 244-245
  analysis of, 246-247, 248-249
  App Store. *See* App Store
  approval, 328-329
  compiling, 71
  crashes, 292
  exiting, 192-193, 229
  full-screen, 76
  home path, 196
  icons, 278-279, 306-309, 312
  images. *See* images
  installing, 66-71
  interface standardization, 74-75
  listing of, 68, 69, 70
  mail, 232-233
  maps, 234-235
  memory consumption over time, 296
  messages, 236-237
  number of, 74
  output settings, 56-57
  profiling, 294
  publishing, 63
  review process, 328
  sales tracking, 330-331
  single task focus, 75
  startup, 209

states, 194–195
   testing, 63, 109
   third-party, 74, 192, 194, 278
   updating, 67, 197
approval, applications, 10, 328–329
.artwork files, 129
AS1/2/3. *See* ActionScript
.as files, 32, 34, 35
AsBreakpoints.xml file, 285
ascent speed, 217
asynchronous handling of files, 206
attachments, 233
Audacity, 131
audio
   challenges, 81
   codec options, 132–133
   importing, 130–131
   relative paths, 134
   sounds, 134–141, 144–145
Auto Orientation, 80
automation, file conversion, 71

B

background color, 52, 596
backing up, 66, 197
Barnes & Noble nook, 171
battery life, 228
bidirectional text, 170
binary file formats, 201
bitmaps as fill, 49
blind carbon copy (bcc), 232
blue, green, red (BGR), 113
blue overlay, 298
brands/branding, 76, 329
breakpoints, 284–285
browsers, 7, 287
buffering, 155–156
bugs, reporting/fixes, 69, 73, 277
builds, testing, 73
bulk uploading, 23
bundle identifier suffix, 24
bundle seed ID, 24
bundling, 118–119, 134–135, 136, 152–153, 208
business location searches, 215
buttons, 38–39, 88–89
buzzwords, 328, 329
`ByteArray`, 199, 201

C

`cacheAsBitmap`, 254–255
`cacheAsBitmapMatrix`, 256–257
cameras, 6, 8, 124, 125
carbon copy (cc), 232
certificate signing requests (CSR), 12–13, 14–15, 16–17, 314
certificates, 18–21, 58–59, 314–315
channels, sound, 145
classes, 30, 34
code
   cleanup, 193
   collapsing, 30
   color effects, 40
   errors, 63
   hint windows, 31
   `MovieClip`, 37
   in objects, 42
   organization, 32
   properties, 41
   for video, 155
   for YouTube, 239
codecs, 132–133, 146
color
   in advertising, 249
   background, 255
   BGR/RGB, 113
   blending, 298, 299
   effects, 40
   icons, 307
   Kuler panel, 93
   status bar, 86
columns (databases), 182, 183, 189
command lines, compiling from, 62–63, 71, 319
commands, 31
comments, 30, 31
compiling, 62–63, 64, 71, 208
compression, 112, 148. *See also* .zip files
computers, testing on, 62
configuration
   compiling, 64
   settings, 54–55, 78
configuration utilities
   about, 5
   bulk uploading, 23
   as information source, 29
   installing with, 68–69
   with Mac OS X, 28, 68

INDEX

provisioning profile installation, 28
  with Windows, 28, 68
constructors, 32
contacts, 218–223
content
  digital, 171
  layout, 84–85
  panning, 98–99
  rotation, 80, 106, 107
cookies, 7, 174
copying files, 208–209
Core Animation instrument, 298–299
CPU with video, 147
crashing applications/reports, 292–293, 328
credentials, 168, 242
cropping video, 149
CS5 Help, 50–51, 53
CSR (certificate signing requests), 12–13, 14–15, 16–17, 314
.csr files, 14
cue points, 149
cutoff frequency, 110, 111
cyan overlay, 298

D

data
  for device sensors, 211
  filtering, 110–111
  saving, 193, 194–195
  from shared objects, 178–179
  in tables, 184–191
databases, 180–181, 201, 208, 218
DatagramSocket class, 8
debugging
  applications, 22
  breakpoints in, 284–285
  with Debug Console, 288–289
  with Flash Debugger, 286–287
  remote, 282
  tools for, 11
  with Variables panel, 291
decoders, 132–133, 146
deleting, 177, 190–191
DER (distinguished encoding rules), 20
descent speed, 217
development, 4–5, 10–11, 14, 16, 58

devices. *See also specific applications; specific devices*
  adding, 22–23
  data from sensors, 211
  listing of, 70
  operating system variations, 258–259
  uploading, 23
  variations, 258
digital signatures, 18, 20, 208
directories, 196–197
`dispatchEvent()`, 47
displaying
  current map location, 212
  fonts, 162–163
  images, 116–117
  lists, 42, 250–251
distinguished encoding rules (DER), 20
distribution
  ad hoc, 318–319
  certificates, 19, 21, 314–315
  profiles, 61
  publishing for, 322–323
downloading, 154, 156, 212, 315
drawing API, 48–49
driving directions, 235
driving games, 102
dual threshold buffering, 156
dynamic images, 122–123
dynamic sound tracks, 139

E

Easter eggs, 102
ECMAScript, 117
editing, 26, 30, 40–41, 115, 320
efficiency
  code hinting, 30
  with Configuration Utility, 68
  monitoring, 300
  programming, 30
  testing, 62, 66, 104
email, phone numbers in, 230
embedding, 146, 150–151, 164–165
encoding. *See* code
encryption, 13, 15
engineering consultations, 17
enterprise environment, 10, 68
error messages, 14, 63, 239, 327, 329
events, 46–47, 149
exiting applications, 192–193, 229

F

Facebook, 311
favorites, phone numbers, 222–223
files
  access order, 303
  appending, 204–205
  bundling, 208
  copying, 208–209
  formats, 201
  location, 329
  reading, 200–201
  reference, 196–197
  size, 112, 114, 203
  synchronous handling, 206–207
  updating, 202–203
  writing, 198–199
filtering, 40, 131
fixed-width fonts, 162
.fla files, 4, 34, 35, 52, 53, 55
flagging moments in time, 294
Flash
  Builder, 4, 32
  cookies, 174
  CS5, 4, 50–51, 53, 62–63
  Debugger, 286–287
  Lite versions, 248, 258
  Media Streaming Server, 153
  Player, 2
  Tracer, 287
Flash Video format (FLV), 7, 30, 148
Flex Builder. *See* Flash Builder
flickering images, 120
focus group testing, 75
fonts, 162–166
formatting, 30
frames, 38, 52
frequently asked questions, 17
full-screen applications, 76, 78–79
functionality information, 272

G

games, 78, 102, 195
GeoLocation, 6
GestureTransformEvent, 7
glare, 306, 308–309
global path settings, 35

global sound control, 143
Google Analytics, 246–247
Google Maps API, 212, 234
GPS, 2, 210–211, 216
GPU hardware, 300
graphics. *See* images
gravity, 110, 111
green overlay, 299

H

H.264 video format, 146, 239
hardware failures, 19, 21
hardware-assisted decoders, 132
help files, 50–51
hiding status bars, 76, 78, 79
high-pass filter, 110
hit states, 82–83
hosting, 153, 224, 225
HTMLLoader, 8
human interface guidelines, 74–75
hyperlinks, 230, 237, 238, 240, 241

I

icons, 278–279, 306–309, 312
ID3 tags, 137
images
  as binary file formats, 201
  bundling, 118–119
  creating, 312–313
  defaults, 128–129
  displaying, 116–117
  dynamic creation, 122–123
  importing, 114–115
  photo library, 124–125
  preparing, 112–113
  at runtime, 120–121
  saving, 199
importing, 114–115, 130–131, 150. *See also specific applications*
In App Purchase, 9
indexing databases, 191
Influxis (streaming video host), 153
Info.plist file, 78, 86
information updating, 75
initialization, 32
input text fields, 166–167

INDEX

installing, 28–29, 66–69, 70–71, 72
instance names, referencing, 40
Instruments
  about, 4, 294–296
  Activity Monitor, 302–303
  Core Animation, 298–299
  Hardware Acceleration Profiler, 304–305
  ObjectAlloc, 296–297
  OpenGL ES, 300–301
interactivity, 37–39, 46
interface, 67, 73, 74–75, 76, 84
Internet, 121, 224–225
interruptions, 329
IP addresses, 282
.ipa files
  for content viewing, 86
  conversion from, 68, 70, 309
  creating, 56, 57, 63
  iTunes association with, 66
iPad, 3, 77, 171. *See also specific applications; specific functions*
iPhone. *See also specific applications; specific functions*
  3G/3GS, 2, 3, 79, 136, 258–259
  certificate signing requests, 12–15
  Configuration Utility, 5, 28, 29, 68–69
  controls standardization, 9
  development, 5, 10–11
  favorites, phone numbers, 222
  hardware, 2
  memory limitations, 74
  Packager, 2, 9
  program enrollment, 11
  SDKs, 11
  variations, 258
  YouTube limitations, 239
iPods. *See also specific applications; specific functions*
  about, 2, 3
  certificate signing requests, 12–15
  favorites, phone numbers, 222
  identifier retrieval, 22
  memory limitations, 74
  variations, 258
iSales, 331
iTunes
  about, 5
  affiliates, 17, 241
  applications, 61, 66
  backing up, 197
  Connect, 5
  identifier retrieval, 22
  installing with, 66–67
  .ipa files with, 57
  Link Maker, 240
  provisioning profile installation, 28
  Store, 240–241
  version number updating, 72

Java applications, 64
JPEG images, 112
JSFL script, 72

Keyboard-Common.artwork file, 129
keyboards, 75, 167, 169, 289
keychain access, 12, 13, 18, 24
keys to Info.plist file, 86
-keystore switch, 64
keywords, 329
Kuler panel, 93

Landscape mode, 76–77, 80–81
latitude, finding, 213, 216
layout, 84–85
left channel, 145
Leopard, 71
libraries, 68, 114, 118, 124–127
licensing agreements, 164, 212
lines, drawing, 48
links, 230, 237, 238, 240, 241
LinkShare accounts, 241
lists
  applications, 68, 69, 70
  classes, 30
  configuration settings, 78
  contacts, 218–221
  devices, 70
  Instruments, 294
  methods, 30
Lite versions, 9
load monitoring, 302

loading
  images, 120–121
  from photo library, 126–127
  shared objects data, 178–179
  sounds, 136–137
  time requirements for, 114, 119, 310
  video, 154–155
location mapping, 210–215, 235
lock icon, 13, 15
login credentials, 168, 242
longitude finding, 213, 216
low-pass filter, 111

M

Mac OS X. *See also specific applications*
  certificates, 12–13, 18–19, 59, 315
  fonts, 170
  image limitations, 128
  Instruments, 296, 298, 300
  user interface standardization, 74
  utilities, 28, 68
  workflow options, 71
  Xcode, 70
magenta overlay, 298
Mail application, 230, 232–233
mapping locations, 210–215, 234–235
Marigo Holdings, 331
masking items, 250
memory
  clearing, 193
  crashes, 292
  limitations, 74, 293
  removing from, 44
  requirements for, 114, 119
  tracking, 296
Messages application, 236–237
metadata, 137, 155
methods, list of, 30
microphone, 9
mock-ups, 75
mouse down states, 88
`MouseEvent`, 252–253
`MovieClips`, 36–37, 39
MP3 files, 132, 137. *See also* audio
multiple devices, 3, 5, 69, 80
multiple sounds, 133
multiple touches, 92–93

N

naming conventions, 24, 32, 56, 116, 117
navigation cue points, 149
nested display lists, 250
NetConnection, 7
networks, 8, 136–137, 146, 153, 205
nondevelopers, 66, 69
nontext-based file formats, 199

O

ObjectAlloc instrument, 296–297
objects, 30, 36, 40, 42–45
offline application use, 329
On2 VP6 video format, 146, 148
opaque backgrounds, 255
Open Screen Project, 2
open source sales tracking options, 331
OpenGL ES instrument, 300–301
OpenSSL bin directory, 20
OpenSSL Web site, 14
operating systems, 19, 21, 258–259
optimizing display lists, 250–251
ordering database rows, 187
orientation, 76–77, 80–81, 106–107
Other.artwork file, 129
output from Flash, 283
overlapping items, 250

P – Q

P12 certificates, 18–21
.p12 files, 315
panning, 98–99, 161
parsing binary files, 201
passwords, 18, 20, 58, 64, 168–169
paths, setting, 34–35
pausing video, 158, 159
PEM files, 20
performance
  bottlenecks, 300
  challenges, 250, 251, 253
  with hardware acceleration profiling, 304
  with images, 113
  optimizing, 301
permissions, 164, 212
pfi.jar files, 64

INDEX

phone calls/numbers, 222–223, 230–231, 236
photo library, 124–127
Photoshop, 115
pinch gesture, 94
Pixel Bender, 8
platforms, multiple, 91
plug-ins, browser, 287
PNG images
  encoding, 123
  icons, 306, 312
  as optimal choice, 112, 113
  saving, 199
Portrait mode, 76–77, 80–81
ports, monitoring, 224, 225
preferences, 262, 275, 276, 280
private keys, 13, 14, 15, 18
profiles, saving, 55
profiling, 294, 295, 298, 304–305
progress indicators, 121, 125
progressive downloads, 154
projects, creating, 52–53
properties, 40–41, 132, 138
prototyping, 24, 25, 27
provisioning profiles, 26–29, 60–61, 316–318, 320–321
-provisioning-profile switch, 64
public keys, 13, 18
publishing, 54–55, 63, 165, 318–319, 322–323
PunkStar Studios, 331
push notifications, 9

rating applications, 326
RAW files, 132
red, green, blue (RGB), 113
red overlay, 299, 304
reference files, 196–197
referencing instance names, 40
registration points, 36, 96, 97
rejection of applications, 328
relative paths, 54, 118
`removeChild()`, 44, 45
`removeEventListener()`, 46
repetitive sound tracks, 139
resolution, icons/screens, 76–77, 307, 312
reverse-domain naming convention, 24, 32
reviewer contact, 329
RGB (red, green, blue), 113

right channel, 145
rotate gesture, 96–97
rotation, objects, 256
rounded corners, 308, 312
Run browser, 295
runtime, 116, 120–121, 151

sales tracking, 330–331
sandbox, 180, 218, 328, 329
saving, 124–125, 193, 194–195, 199
scaling, 94, 98, 112, 256, 278
screens, 74, 75, 76–77, 80–81, 84
screenshots, 122, 311, 312, 313
Script Navigator, 30
Script pane, 30
scrollable text fields, 172–173
SDKs, 11
secure socket layer (SSL), 13, 14, 15, 20
security
  login credentials, 168, 242
  passwords, 18, 20, 58, 64, 168–169
  social networking sites, 242
  SQL injection attacks, 184
sequencing images, 114
ServerSocket class, 8
settings
  bundle, 260–262, 274–276, 278, 279
  global, 280–281
  group, 262–263
  icons for, 278–279
  multiple value, 266–267
  slider, 270–271
  text, 264–265
  title, 272–273
shapes, drawing, 48
shared objects, 7, 174–179
sizing files, 40, 112, 126, 148, 203
skeleton custom classes, 32–33
Sleep mode, 228, 229
Smaato, 244–245
SMS messages, 236
Snow Leopard, 71
social networking sites, 242
software decoders, 132
Sony Reader, 171
Sorensen Spark video format, 146

Soundbooth, 131
`SoundMixer`, 140, 143, 144
sounds. *See* audio
source paths, 34–35
special characters, 230
speed, determining, 216
splash screens, 310–311
`Sprite`, 37, 48
SQL injection attacks, 184
SQLite, 7, 180–181, 182–191, 208, 218
SSL (secure socket layer), 13, 14, 15, 20
Stage
  dimensions, 52
  images imported to, 114
  objects on, 30, 36, 42–45
  orientation, 6
  scripting in, 30
  symbols on, 40
standard program pricing/features, 10
statements, trace, 282–283, 287
states, applications, 194–195
statistics, 300, 301, 302
status bar, 76, 78, 79, 86–87
-storetype switch, 64
storyboards, 75
streaming video, 146, 153
submissions to App Store, 324–327
.swf files, 35, 54, 64, 72, 151
swipe gesture, 100–101
switches, 64, 268–269
synchronization
  avoiding, 68
  databases, 181
  deletion, 28
  files, 135, 206
  with iTunes, 66
  pitfalls, 56, 67
syntax, 30, 309
system idle mode, 228–229

T

tables, SQLite, 182–191
tablet computers, 171
-target switch, 64, 65
team agents, 320
templates, 52, 74, 128

testing
  builds, 73
  features, 27, 309
  with focus groups, 75
  multiple devices, 3, 69
  by nondevelopers, 61
  by window resizing, 77
testing applications
  ease of, 258
  efficiency, 66
  quick, 65
  requirements for, 10, 26, 104
  steps for, 62–63, 109
text fields, 166–173
text layout framework (TLF), 170–171
text messaging, 236
textures, 304, 305
third-party applications, 74, 192, 194, 278
3G/3GS
  about, 3
  data transmission with, 79
  downloading expenses, 136
  hardware, 2
  performance, 258–259
TIFF files, 312
Timeline, 30, 36, 116
TLF (text layout framework), 170–171
touch events, 7, 90–91
touch-enabled applications, 82–83, 88–89
touches, multiple, 92–93
trace statements, 282–283, 287
trademarked words, 328
transparent backgrounds, 255
travel speed, determining, 216
Twitter, 242–243
typing challenges, 75

U

uninstalling, 68, 69
unique device identifier (UDID), 22, 316, 317
updating
  applications, 67, 197
  automatic, 131
  data, 188–189
  files, 202–203
  preferences, 275, 276

INDEX

screen regions, 298
stored information, 75
  with Variables panel, 291
  version numbers, 56, 72–73
uploading, 16–17, 22, 23, 324–327
URL scheme, 236, 238, 240
usability, 75
user guides, 16
user information storage, 174
user interface. *See* interface

V

vectors, 48, 251
version numbers, 56, 67, 69, 70, 72–73
vertical text, 170
video
  buffering, 155–156
  bundling, 152–153
  controlling, 158–159
  converting, 148–149
  embedding, 150–151
  formatting, 7, 146–147
  loading, 154–155
  panning, 161
  pausing, 158, 159
  playing, 158
  resuming, 158
  stopping, 158, 159
  visual elements, 147
  volume, 160–161
  YouTube, 238–239
visual assets, 250
visual feedback, 82, 88, 209
visualizations of sound spectrum, 144–145
volume, 131, 142–143, 160–161

W

warnings, 63. *See also* error messages
WiFi, 79, 226–227
wildcard App IDs, 25, 27, 60, 320
wildcard bundle identifiers, 24
Windows
  certificates, 14–15, 20–21, 315
  fonts, 170
  utilities, 28, 68
windows, number of, 74
workflow options, 70, 71
writing systems, 170

X

X.509 Base64 encoded distinguished encoding rules (DER) certificates, 20
Xcode, 4, 29, 70–71, 260, 279
.xfl format, 53
.xml files, 55, 72

Y

Yahoo! maps API, 212
yellow overlay, 298
YouTube video, 238–239

Z

.zip files
  for content viewing, 86
  for file conversion, 57, 68, 70, 71, 309
  icons, 312
zoom events, 94–95